Statistical Concepts
in Geography

Statistical Concepts in Geography

JOHN SILK
University of Reading

London
GEORGE ALLEN & UNWIN
Boston   Sydney

First published in 1979
Reprinted in 1981

GEORGE ALLEN & UNWIN LTD
40 Museum Street, London WC1A 1LU

British Library Cataloguing in Publication Data

Silk, John
    Statistical concepts in geography.
    1. Geography — Statistical methods
    I. Title
    519. 5'02'491   G70.3   78—40957

    ISBN 0—04—910065—3
    ISBN 0—04—910066—1 Pbk

Typeset in 10 on 12 point Press Roman by Preface Ltd, Salisbury, Wilts.
and printed in Great Britain
by Unwin Brothers Limited, Old Woking, Surrey

In memory of Philip

Preface

This book is intended primarily for first-year undergraduate students in universities, polytechnics and colleges of further and higher education, and as such it enters what is an increasingly crowded market. However, I believe there is room for a text which tackles a number of particularly important problems at the introductory level. First, basic statistical concepts, such as those of probability, independence, randomness and sampling distribution, are emphasised, and illustrated by in-class experiments supplemented by computer-generated results. In this respect, the overall objective is to introduce greater rigour with little or no increase in mathematical formality. Secondly, the student may work through the text at her or his own pace, checking on the grasp of concepts and techniques by doing the in-text boxed exercises before reaching the more testing territory in the exercises at the end of each chapter. Thirdly, I have tried to show the relevance of statistical techniques to geographic problems through the use of numerous real-world data sets — some of which are taken from student project and thesis work — in the text and in exercises, backed by carefully selected references. Finally, no mathematical competence beyond sound Ordinary Level or high-school graduation standard is assumed, and every effort has been made to minimise the initial feeling of 'symbol shock' that many students encounter. Although complex derivations are avoided, references to more advanced technical treatments are also provided for those interested. Most of the ideas in the book have arisen while teaching a course at Reading for the past nine years and, needless to say, result from numerous trial-and-error exercises which students have not always borne stoically.

Many people helped, directly and indirectly, with this book. Dick Chorley and Peter Haggett inspired me to become an academic geographer in the first place, and all my colleagues at the Department of Geography in Reading University have provided a very pleasant atmosphere in which to work. In particular, Geoff Lucas commented on various parts of an early draft of the manuscript, and Sophie Bowlby, Ian Fenwick, Dave Foot, John Hardy, Trevor Meadows and John Townshend kindly made available sets of data. Ronnie Savigear provided many papers giving examples of statistical applications in physical geography, and I am particularly grateful to all those physical geography colleagues who patiently explained the background to some of the physical geography exercises to a 'mere human'! A number of colleagues in the Department of Applied Statistics at Reading University, particularly Roger Mead, very kindly discussed a number of statistical issues with me at some length, although of course all responsibility for the particular interpretations presented, and any errors, are mine alone. I am also extremely grateful to Angela

Gurnell, Dave Unwin and Neil Wrigley for providing detailed comments on the 'penultimate draft' — they caused me an awful lot of hard work which I hope has resulted in a marked improvement in the organisation and quality of the text. Thanks also to Joyce Gillo, Patricia Hobson and Linda Tarrant for typing the earlier drafts, and to Debbie Lewis for producing an excellent final manuscript. My wife Cathy said there was no reason why I should not be capable of producing this book, and she was right.

John Silk
Reading, January 1979

Contents

I

INTRODUCTION TO STATISTICS IN GEOGRAPHY

1 Introduction

Statistical methods are used by geographers because they help us to come to conclusions based upon **empirical** data, these being measurements derived either from observation or experiment. Geography was a strongly empirical subject well before the 'quantitative revolution' of the 1960s, and so statistical methods can be regarded as essential aids to geographic enquiry.

The way in which empirical evidence and conclusions are related statistically can be illustrated if we consider briefly the difference between mathematics, on the one hand, and statistics, on the other. Mathematics is chiefly **deductive** in nature. Consider the following statements:

(1) A is greater than $B (A > B)$, and
(2) B is greater than $C (B > C)$.

Provided both statements or premises are true, it follows logically that:

(3) A is greater than $C (A > C)$.

We can be quite certain that the conclusion expressed in statement (3) is correct, given the premises expressed in statements (1) and (2), just as we can be quite certain that the theorems of Euclidean geometry are correct, and such logical necessity is a hallmark of deductive reasoning. No matter how complex the argument, therefore, no additional information is required to reach a firm conclusion. In this sense, deductive arguments cannot go beyond the information given.

The position is different in statistics, which is primarily **inductive** in nature. Inductive arguments provide conclusions which in some sense exceed the content of the premises upon which they are based. Suppose we polled a representative sample of adults living in a suburb of a large city, and found that 20% of them commuted to work in the city centre each day. This may be stated as a premise:

(a) 20% of the adults in the sample from the suburb commute to the city centre.

From this, it might be concluded that:

(b) 20% of *all* adults in the suburb commute to the city centre.

However, we should feel bound to add a qualification to the effect that the conclusion is 'probably true' or 'approximately correct', simply because not all adults were questioned. Because of the gap in our knowledge, it is necessary to infer that approximately 20% of all the adults are commuters. This belief or opinion is the result of an **inductive inference**. A **statistical inference** is a form of inductive inference which allows the investigator to be relatively precise about her or his degree of uncertainty, stating, for example, that there is a 95% chance or probability that the true percentage of adults commuting to the city centre lies between the limits set by 14% and 26%. The techniques of **inferential statistics** provide formal procedures for calculating such limits and probabilities, for testing statistical hypotheses, and thereby drawing statistical inferences.

The overall plan of the book is shown in Figure 1.1. Although Chapter 2, on descriptive statistics, is primarily concerned with methods for summarising characteristics of large bodies of data, it should be clear that informal inferences must be made leading to interpretations based on the investigator's own knowledge and judgement. This also holds true for all ensuing chapters. Following this (Ch. 3), we describe two contexts in which inferential statistics are generally

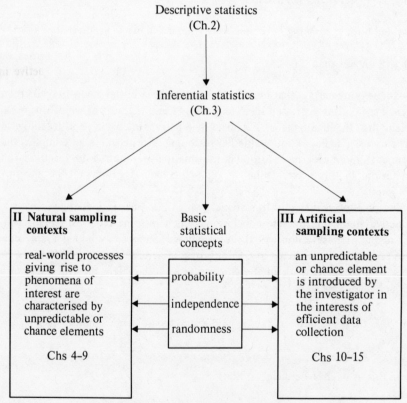

Figure 1.1   Overall plan of the book.

employed: **natural sampling contexts**, involving study of real-world processes giving rise to phenomena characterised by unpredictable or chance elements, and **artificial sampling contexts**, in which an unpredictable or chance element is introduced by the investigator in the interests of efficient data collection. Sections II and III are devoted to techniques employed in natural and artificial sampling contexts respectively, although many techniques can be used in either context, as is made clear where appropriate in the text. The basic statistical concepts of probability, independence and randomness are discussed in detail in Chapters 4 and 5, and reappear throughout the book, particularly in Section II and early in Section III.

Brief comments should be made on the quality of information or data upon which any statistical analysis is based. The data should be both valid and reliable. Measurements are said to be **valid** if they really measure what we think they are measuring – if we are trying to ascertain an individual's knowledge of places in her or his home town, then the number of correct identifications may be regarded as a more valid measure if photographs of places are presented than if place names or maps are used. This is simply because use of names or maps is liable to provide a better measure of an individual's ability to remember names or road maps than of their spatial knowledge. Measurements are said to be **reliable** if they are free from substantial bias – poorly maintained equipment or careless handling of soil or vegetation samples may lead to consistent over- or underestimates of cation exchange capacities or species counts – and liable to relatively small errors. Official statistics should never be regarded as sacrosanct in this respect – errors of up to 22% have been reported as a result of checks carried out on the 1966 Sample Census of England and Wales, and errors exceeding 5% were quite common (Gray & Gee 1972). Careful choice and correct use of a statistical technique counts for little if the measurements are of poor quality – as the computer experts say 'garbage in means garbage out'!

REFERENCES AND FURTHER READING

Comments on census errors may be found in:
> Gray, P. and F. A. Gee 1972. *A quality check on the 1966 ten per cent sample census of England and Wales.* London: H.M.S.O., Office of Population Censuses and Surveys, Social Survey Division.

A discussion of data reliability and validity may be found in:
> Nachmias, D. and C. Nachmias 1976. *Research methods in the social sciences.* London: Edward Arnold.

An introductory treatment of the topics of inference, deduction and induction is given in:
> Salmon, W. C. 1963. *Logic.* Englewood Cliffs: Prentice-Hall. (Chs 1–3).

2 Descriptive Statistics

INTRODUCTION

In this chapter we first consider the raw materials with which descriptive and inferential statistical techniques work — measurements or data. Methods for summarising large bodies of data are then described, making reference both to devices such as tables, charts and graphs, and to statistical measures of central tendency and dispersion. The final section deals with statistics for summarising certain kinds of spatial data.

MEASUREMENT

The term 'measurement' may refer both to an activity (or process) and to an outcome (or product) of that process. It is convenient to consider these two aspects of measurement in reverse order.

Measurement as **product** may be considered in terms of four levels or scales (Stevens 1946). The most rudimentary level is provided by the nominal scale, allowing us to classify or categorise phenomena. Households may be placed in different social groups, or pebbles classed according to the rock type from which they are derived. Regional classifications provide an important example of nominal scale measurement in geography. Binary data represent a special case whereby phenomena are placed in one of only two possible classes, e.g. female or male, on or off a sandstone outcrop. The categories employed in constructing our nominal scale should be **exhaustive**, including all phenomena under study, and **mutually exclusive**, so that no item can be placed in more than one category. What we can say about any two items measured on such a scale is that those in the same category are equal, and those in different categories unequal or different, in terms of the characteristics used to define the categories.

Measurement at the ordinal scale is achieved if there is enough information to place the categories in rank order. Stream segments or central places may be assigned to one of several classes arranged in hierarchical order. Such measurement is said to provide a **weak ordering**, because we can use the terms 'more than' or 'less than' with respect to two items in different classes, but not for two items in the same class. If each item is ranked, as is the case with areas placed in order of residential desirability by an individual, or rocks ranked according to Moh's scale of mineral hardness, then a **complete ordering** is achieved. Even if such an ordering is possible, the quantitative difference between each pair of ranked items remains unknown. However, an ordinal scale assigns numbers – the most preferred area receives rank 1, the next most preferred rank 2, ...

Both the nominal and weakly ordered ordinal scales yield **count data**, because the number of items falling in each category may be determined.

The³ _interval scale_ assigns an exact numerical value so that the difference between any two items on the scale is known precisely. If the midday temperatures on successive days were 16 and 20 °C, then the difference is 4 °C. Interval scales lack true zero points – for instance 32 °F is equivalent to 0 °C. Psychological scaling methods for measuring attitudes and preferences may also give measurements at this level.

The maximum amount of information is provided by the⁴ **ratio scale**. All such scales possess a true zero point, as well as permitting precise differences to be calculated, so that measurements retain the same ratio to one another no matter what units are employed. For example, distances of 50 and 100 miles form the same ratio (½ or 2) as their equivalents 80·5 and 161·0 in kilometres. This is not true for, say, temperatures.

It is sometimes helpful to reduce measurements at a high level to those at a lower level by combining age or income data into classes, or locations by distance bands from a shoreline or city centre.

Some interval or ratio measurements are based on a **continuous** scale. Provided we are prepared to pay for the necessary instruments, distance, weight and elevation measurements can be determined to any desired level of accuracy. Location can also be measured on such a scale, as shown by the 'northings' and 'eastings' obtainable from a topographic map. Continuous measurements can take on any value within certain limits, e.g. greater than or equal to zero for distances or weights, any value between plus and minus infinity (in theory) for slopes. **Discrete** measurements are in whole units or **integers** (including zero) – there can only be 1, 2, 3, ... persons or rooms in a household, although of course the average number of persons per household in a census tract or ward may be 2·91 and measured on a continuous scale.

As **process**, measurement provides the essential link between abstract concepts, not in themselves observable, and measurements. Concepts may be regarded as abstractions or generalisations in the sense that they may be applied to observations of a wide variety of phenomena. For example, we use 'temperature' as a concept which refers to the relative 'hotness' or 'coldness' of phenomena, and 'social class' to refer to similarities between characteristics of people or households in terms of their attitudes, perceptions and behaviour. For many purposes, real-world phenomena may be discussed adequately in these terms. However, it is impossible to proceed with any statistical analysis unless some means is found of expressing a concept in terms of one of the measurement scales already described. This is provided by an **operational definition,** which specifies the measurement process in terms of the methods or 'operations' required to obtain measurements (Fig. 2.1). These may be quite elaborate. For instance, measurement of 'daily temperature' poses problems because at meteorological stations such measurements are read off in degrees Fahrenheit or Centi-

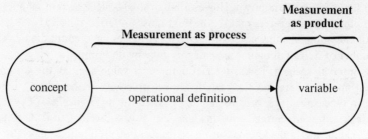

Figure 2.1   Relationship between concept and variable.

grade from a continuous graph traced out on a chart fixed to a revolving drum. Which of the infinitely large number of instants during the day (i.e. the period from midnight to midnight — this in itself representing an operational definition of the term 'day') provides the desired measurement? We might define daily temperature as 'temperature at midday' or as the average of the maximum and minimum temperatures occurring during the 24-hour period. Operational definitions of 'social class' for households are frequently based on the occupation of the so-called 'head of the household' as recorded in census returns or questionnaires. Lists have been published giving the category, e.g. professional, semi-skilled (see Table 2.1), into which any occupation may be placed (*Classification of occupations* 1970).

Operational definitions rarely, if ever, provide a 'pure' measure of the concepts to which they refer, giving rise to problems of reliability and validity as mentioned in the Introduction. This is because the information is difficult or expensive to obtain. Social classes defined by occupation will correspond only crudely to those based on a concept expressed in terms of similarities of attitudes, perceptions and behaviour. A relatively cheap and convenient classification is provided, nevertheless, and greater time and expense can be invested in developing a sophisticated categorisation if warranted.

> What difficulties do you foresee in providing operational definitions of household, accessibility, slope angle and atmospheric pressure, or of basic geographical concepts such as distance, direction and shape (Taylor 1977, pp. 42–4, 49–52)?

Further discussion of formal properties of measurements (measurement as product) may be found in Taylor (1977, pp. 38–41) and of measurement as process in Taylor (1977, pp. 41–65) and Nachmias and Nachmias (1976, pp. 15–20). In the remainder of this chapter, measurement is considered almost entirely as product.

VARIABLES AND OBSERVATIONS

A **variable** is a measurable property of any person, place or other phenomenon of interest, and so refers to the products of measurement. In a survey of two neighbourhoods in Crawley New Town, England (Langley Green and Gossops Green) the property of interest in Table 2.1 is social class. Each household is measured on a nominal scale of five categories. Often, a variable is represented by the symbol X, e.g. X = social class, X = temperature.

An **observation** is the value assigned to each item under investigation, and is represented by X_i (X subscript i). If the total number of items is n, then i ranges from 1 to n. In the Langley Green case, $n = 163$. If the 23rd household interviewed in the sample can be classified as 'unskilled manual', then X_{23} is placed in that category. Observations at a higher scale of measurement are shown in Table 2.2, and represent annual rainfall in centimetres for the city of Kano in Nigeria.

> What scale of measurement do the Kano observations attain? Is this variable discrete or continuous? How is annual rainfall defined operationally?

SUMMARISING DATA

The measurement process may yield an extremely large number of observations, perhaps for several variables. Considering only one variable, there would be $n = 163$ observations on social class for Langley Green, and $n = 68$ observations on rainfall for Kano. Such bodies of data are difficult to comprehend or compare unless summarised using devices such as tables, charts and graphs, or statistical measures of central tendency and dispersion.

The social class observations on sample households in Langley Green and Gossops Green may be recorded on a **tally sheet** (Table 2.1a). The set of counts or frequencies obtained from the tally are given in Table 2.1b and represent the **frequency distributions** of households by social class for the two areas.

> Compare the 'class structure' of the two areas by converting the frequency in each category into a percentage.

Observations at higher scales of measurement may also be presented as frequency distributions. The Kano rainfall measurements of Table 2.2a are for the 70-year period 1905–74 (observations for 1967 and 1968 are missing). Because there are no 'natural classes' into which observations may be placed, a

Table 2.1   Class structure of two Crawley New Town (UK) neighbourhoods (1961)

(a) Results of a survey of two districts that classified people into one of five social classes.

Class	Langley Green	Gossops Green
I Professional	11	111 11111
II Intermediate professional	11 11111 11111	111 11111 11111 11111 11111
III Unskilled non-manual; Skilled manual	11111 11111 11111 11111 11111 11111 11111 11111 11111 11111 11111 11111 11111 11111 111	11111 11111 11111 11111 11111 11111 11111 11111 11111 11111 11111 11111 11111 11111 11
IV Semi-skilled	11111 11111 11111 11111 11111 11111	111
V Unskilled manual	11111 1	11111 1

(b) Frequency distributions for raw data in (a).

Class	Langley Green	Gossops Green
I	2	8
II	12	20
III	113	62
IV	30	3
V	6	6
Totals	163	99

Percentage calculation

$$\text{percentage in class} = \frac{\text{number in class}}{\text{total number in area}} \times 100$$

$$\text{percentage in class IV (Langley Green)} = \frac{30}{163} \times 100 = 18.4\%.$$

(c) Histograms of data.

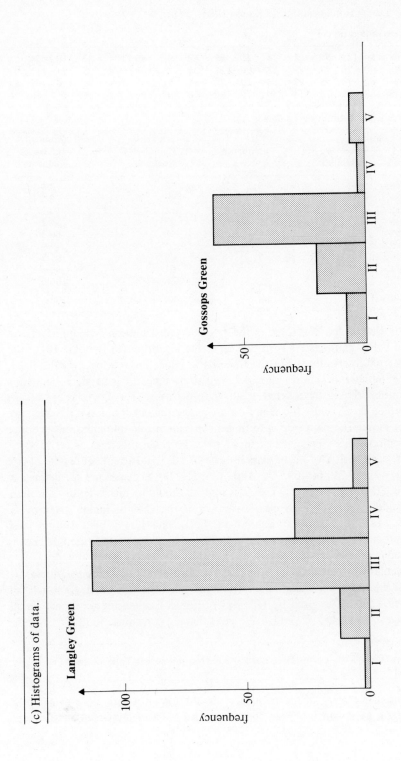

Table 2.2   Annual rainfall at Kano (1905–74).

(a) Observations (in cm).

97	65	89	123	68	101	74	48	69	82	99	86	104	82	110	54	93
78	85	90	70	77	89	83	100	118	101	85	83	94	98	74	111	91
81	84	62	79	48	99	106	80	70	58	92	81	104	71	110	107	84
101	83	102	76	78	114	70	75	90	78	97	96	71	66	41	56	76

(b) Frequency distribution for data in (a).

Group		Frequency f_j	Cumulative frequency	Cumulative percentage frequency
Limits	Midpoint			
40–49	45	3	3	4·4
50–59	55	3	6	8·8
60–69	65	5	11	16·2
70–79	75	15	26	38·2
80–89	85	15	41	60·3
90–99	95	12	53	77·9
100–109	105	9	62	91·2
110–119	115	5	67	98·5
120–129	125	1	68	100·0
Total		68	68	100·0

rule of thumb for constructing the frequency distribution is required. There should be no more classes than five times the logarithm (to the base 10) of the number of observations. For the Kano data, $5(\log_{10} n) = 5(\log_{10}(68)) \approx 9·15$, so the number of classes should be eight or nine. In general, a frequency distribution should have between six and 16 classes. Too few classes obscure interesting detail, while too many do not repay the effort unless n is 'large', say 500 or more. The next step is to find the maximum and minimum values of the observations, those being 123 and 41 here. To cover this range, a set of $k = 9$ classes, each with a width or **class interval** of 10, is sufficient. Thus, the classes are 40–49, 50–59, 60–69, . . . , 120–129. Notice the classes are *not* defined as 40–50, 50–60, etc (why?). The class interval should be the same for each class. The middle value of each class is known as the **class midpoint**, given by 5, 15, . . . , 115, 125 for the Kano data. Such values may be useful in later calculations. Table 2.2b shows how the frequency in the jth group or class is denoted f_j, so that $f_1 = 3, f_2 = 3, . . . , f_9 = 1$.

Cumulative frequencies give the number of observations whose measurements do not exceed a particular value. The number of years in which Kano rainfall was below 60 cm is given by $(f_1 + f_2) = 6$. Such frequencies are shown for all nine classes in Table 2.2b, which also gives them in percentage form.

> Check the cumulative percentage frequencies of Table 2.2b.

A visual summary of the frequency distribution information is provided by a **histogram**, as shown by those constructed for the Langley Green and Gossops

(a) Frequency polygon

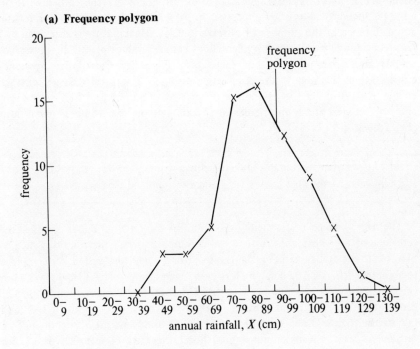

(b) Cumulative frequency curve (%)

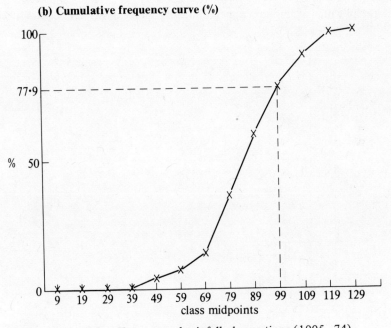

Figure 2.2   Kano annual rainfall observations (1905−74).

Green social structures in Table 2.1c. Two histograms may be compared if the units on the vertical axis are expressed in percentages, because this corrects for any difference in the number of observations. A visual picture is also given by a **frequency polygon**, in which straight lines connect values placed at appropriate heights above the midpoints of each class (Fig. 2.2a). **Cumulative frequency curves** extend this idea to the cumulative case and, if expressed in percentages, enable us to read off the percentage of years with rainfall less than a given level and, by subtraction, the percentage with rainfall exceeding a given level (Fig. 2.2b).

> Construct histograms showing the percentage of households in each of the five classes for Langley Green and Gossops Green.

The overall shape of frequency distributions can be described in terms of the concepts of skewness and kurtosis. A **skewed** frequency distribution, such as that of the October rainfall figures for Kano over the period 1905–74, is asymmetric (Fig. 2.3). The majority of measurements fall below 20 mm (measurements are in mm because the October totals are rather low). This distribution shows extreme **positive** or **right skewness**, with a long 'tail' extending to the right. Other examples of such distributions are those of income, settlement size and maximum annual flood. **Negative** or **left skewness** is less common, showing a long 'tail' to the left, and may occur where percentage data tend to 'pile up' against an upper limit, e.g. percentage of households with

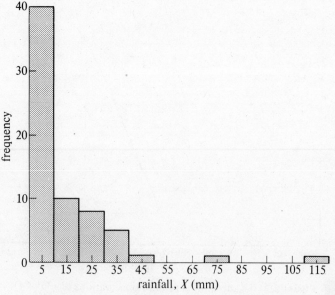

Figure 2.3   October rainfall at Kano (1905–74).

television sets in each county in England. A symmetrical distribution has no skewness.

> Use the frequency distribution of Table 2.2b to construct a histogram for the Kano annual rainfall data. Does the distribution show any sign of skewness and, if so, of what kind?

Not important

Kurtosis refers to the degree of peakedness of a frequency distribution, but the concept is not very important unless detailed comparison with the theoretical normal distribution (Ch. 7) is required.

> Compare the peakedness of the histograms for October and annual rainfall at Kano. Should the concepts of skewness and kurtosis be applied to measurements at the nominal scale?

For calculations of measures of skewness and kurtosis, see Ebdon (1977, pp. 26–9) or Norcliffe (1977, pp. 49–53).

Statistical measures

These measures fall into two major groups — measures of **central tendency** and measures of **dispersion**. The former give an idea of the average or 'typical' value of a set of observations, whereas the latter are concerned with the 'spread' or degree of variability in a set of measurements.

The **mode** is the simplest and most readily calculable measure of central tendency, taking on the 'value' associated with the class containing the highest number of observations. For the social structure data, the modal class for both neighbourhoods is class III (unskilled non-manual and skilled manual), and for the rainfall data it is the 80–89 cm class, so we take the class midpoint of 85 cm as the mode. A frequency distribution with a single peak is called **unimodal**, those with two peaks **bimodal**. The mode is particularly useful for describing the most typical value or class of a skewed distribution.

The **median** represents the central point of a set of ranked observations. If the number of observations n is even, the median value falls midway between the two central ranks. For the Kano data, ranked in Figure 2.4, the 34th and 35th observations have values of 83 and 84 cm, so the median is $(83 + 84)/2 = 83 \cdot 5$. For an odd number of observations, the median takes on the value of the central rank, e.g. if $n = 21$, the median value is that of the observation ranked 11. Summary measures of skewed variables such as income, or annual rainfall totals in arid areas, are often reported as median values. Notice that the median cannot be calculated for nominal data — it makes no sense to talk of the 'median social class'!

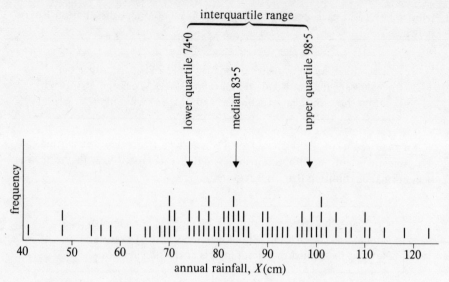

Figure 2.4   Dispersion graph of annual rainfall at Kano (1905–74).

The **arithmetic mean** – usually called the **mean** – is the most commonly used measure of central tendency and is the mathematical name for the average of a set of observations. The symbol for the mean is $\bar{X}$, where X represents any variable of interest and

$$\bar{X} = \frac{X_1 + X_2 + X_3 + \ldots + X_n}{n} = \frac{1}{n}\sum_{i=1}^{n} X_i \qquad (2.1)$$

The symbol Σ is the Greek capital letter sigma and tells us to add up the n observations on variable X. The subscript i can take on any value from 1 to n. For the Kano annual rainfall data, X = annual rainfall, and mean annual rainfall is given by

$$\bar{X} = \frac{1}{68}\sum_{i=1}^{68} X_i = \frac{X_1 + X_2 + \ldots + X_{68}}{68}$$

which, from Table 2.2a gives

$$\bar{X} = \frac{93 + 97 + 65 + \ldots + 66 + 41 + 56}{68} = \frac{5757}{68} = 84\cdot66$$

(Because it is understood that the summation is over all n observations, the symbols above and below the Σ sign may be omitted. Unless otherwise stated, this is the assumption made in this text.)

The mode, median and mean of a perfectly symmetrical frequency distribution are identical. For such distributions, therefore, any of these measures is as good as any other for many purposes, although the mean is particularly important in inferential statistics.

It is better to use the median, or the mode, as a measure of central tendency for skewed distributions because neither is as sensitive as the mean to a small number of extremely high or low values of X. For the October rainfall observations at Kano, the mode is 5·00 mm and the median is 5·05 mm, but $\bar{X} = 12·44$ mm, influenced by the extreme values in the 'tail' of the distribution (Fig. 2.3). Omitting the extreme observation of 114·6 mm for 1932 reduces the mean from 12·44 to 10·92 mm, but leaves the mode unaltered and changes the median from 5·05 to 5·00 mm.

> Compare these three measures of central tendency for Kano annual rainfall. What do you conclude?

A **proportion**, denoted p, is a special kind of arithmetic mean. If p is the proportion of households in Langley Green in social class III, $X_i = 1$ if a household is in class III and $X_i = 0$ otherwise, then, taking $n = 163$ (the number of households from Langley Green), we have

$$\Sigma X_i = (113 \times 1) + (50 \times 0) = 113$$

and

$$p = \bar{X} = \frac{1}{163} \Sigma X_i = \frac{1}{163} \times 113 = 0·693$$

The mean may be calculated rapidly from frequency distribution information, rather than from individual observations. If C_j is the midpoint of the jth class, and f_j is the number of observations in the jth class, then we have

$$\bar{X} = \frac{1}{n} \sum_{j=1}^{k} f_j C_j \tag{2.2}$$

where k is the number of classes. For the annual rainfall data in Table 2.2b this gives

$$\bar{X} = \frac{1}{68} [(3 \times 45) + (3 \times 55) + \ldots + (5 \times 115) + (1 \times 125)] = \frac{5810}{68} = 85·44$$

an answer differing by only 0·78 from that obtained earlier. Use of Equation 2.2 assumes that observations are evenly distributed within each class. This may not always be the case.

> Construct a frequency distribution of the October rainfall observations from the histogram in Figure 2.3, and then use equation 2.2 to calculate $\bar{X}$ (in mm). How does the result compare with that of 12·44 mm previously obtained? How would you account for any difference?

Measures of dispersion should also be used, either because distributions with the same mean differ markedly in 'spread' around this value, or because the 'spread' in any distribution may be large relative to the mean.

The simplest measure of dispersion is the range, given by the difference between the largest and smallest observations. For most statistical purposes it is too sensitive to extreme values to be of use. However, it is valuable in some areas involving design, such as construction of dams, irrigation and flood-control schemes in the form of differences between maximum and minimum river discharge or the highest and lowest tides.

Measures such as percentiles and the **interquartile range** are associated with ranked data and the median. In accessibility studies, it is useful to know the distance within which a specified percentage of households are located from workplaces or other facilities. Guy (1976) found that 60% of 657 sample households in Reading were located within 800 m (walking distance) of convenience goods outlets, i.e. the 60th percentile value is 800 m.

A measure of spread about the median, the interquartile range, is given by the difference between the 25th and 75th percentile values, these being the values below which 25% and 75% of the observations fall respectively. For the Kano annual rainfall data, these values lie between the 17th and 18th and 51st and 52nd observations (lowest value is ranked first) and are 74·0 and 98·5 (Fig. 2.4). This gives an interquartile range of (98·5 − 74·0) = 24·5. Quartiles are often used to compare precipitation and temperature records from different stations. Observations may be divided into any kind of **quantile**, e.g. terciles into three equal sets, quintiles into five equal sets.

> Find the 25th and 75th percentile values for the Kano October rainfall figures given in Table 2.3, and construct a chart showing these values and the interquartile range. Compare your findings with those for annual rainfall shown in Figure 2.2.

The variance and standard deviation are very important statistical measures of dispersion, partly because each has desirable properties in the realm of inferential statistics. Both are based on interval or ratio data. The **variance** is obtained by subtracting each observation from the mean $\bar{X}$, squaring the resulting differences $(X_i - \bar{X})$ to eliminate minus quantities, adding them up to give the sum of squares $\Sigma(X_i - \bar{X})^2$, and finally dividing by the number of observations n. The resulting **mean square** is denoted by σ^2 (sigma squared) where σ is the Greek

Table 2.3   October rainfall at Kano (1905−74) (values in mm).

28·4	0·3	0·5	19·6	0·0	27·2	6·1	5·1	7·1	4·6	2·0	0·0	1·0	0·0	3·0	26·9
7·9	16·0	18·3	1·3	1·0	42·4	12·7	24·1	2·3	114·6	0·0	1·8	1·0	0·0	0·0	22·9
30·5	0·0	0·0	5·6	14·2	28·7	13·2	1·3	0·0	0·0	9·7	11·7	24·1	1·8	34·3	32·8
12·2	0·0	1·8	0·5	0·0	37·8	0·0	0·0	11·7	0·0	70·1	3·0	0·0	5·1	0·0	5·0

lower case letter sigma. In symbols, we have

$$\sigma^2 = \frac{1}{n} \Sigma (X_i - \bar{X})^2 \qquad (2.3)$$

Using this **definitional formula**, columns (1), (2) and (3) of Table 2.4 show how values should be set out for calculations on the Kano annual rainfall data, and we find $\sigma^2 = 300 \cdot 75$. However, with a little algebra, Equation 2.3 reduces to the **computational formula**

$$\sigma^2 = \frac{1}{n} \Sigma X_i^2 - \bar{X}^2 \qquad (2.4)$$

requiring fewer and easier calculations.

Show that using the computational formula gives $\sigma^2 = 301 \cdot 05$.

The most important measure of dispersion based on interval or ratio data is the **standard deviation**, σ, calculated simply by taking the square root of the variance, so that

$$\sigma = \sqrt{\frac{1}{n} \Sigma X_i^2 - \bar{X}^2}. \qquad (2.5)$$

[Computational formula]

For the annual rainfall data, $\sigma = \sqrt{301 \cdot 05} = 17 \cdot 35$ by this formula.

Table 2.4 Quantities required for calculation of variance and standard deviation for Kano annual rainfall data.

X_i	$X_i - \bar{X}$	$(X_i - \bar{X})^2$	X_i^2
93	8·34	69·56	8649
97	12·34	152·28	9409
65	−19·66	386·52	4225
89	4·34	18·84	7921
123	38·34	1469·96	15129
⋮	⋮	⋮	⋮
96	11·34	128·60	9216
71	−13·66	186·60	5041
66	−18·66	348·20	4356
41	−43·66	1906·20	1681
56	−28·66	821·40	3136
Totals 5757	−	20451·22	507849

Definitional formula

$$\sigma^2 = \frac{1}{n} \Sigma (X_i - \bar{X})^2 = \frac{1}{68} \times 20451 \cdot 22 = 300 \cdot 75$$

$$\sigma = \sqrt{300 \cdot 75} = 17 \cdot 34.$$

The variance and standard deviation may also be obtained from a grouped frequency distribution using the equation:

$$\sigma^2 = \frac{1}{n} \left(\sum_{j=1}^{k} f_j C_j^2 - \left[\sum_{j=1}^{k} f_j C_j \right]^2 \Big/ n \right) \tag{2.6}$$

where C_j is the midpoint of the jth class, f_j is the frequency in that class, and k is the number of classes. We then calculate σ as before.

The variance is particularly sensitive to extreme observations, and the standard deviation to a lesser extent (because of the square root operation). This is very noticeable if distributions are markedly skewed. Under these conditions, the interquartile range is to be preferred as a measure of dispersion about a central value.

The **coefficient of variation** (CV) is helpful when comparing observations for different places or times, even if they are expressed in different units of measurement, because there tends to be a strong relationship between the *absolute* size of the observations (as represented by their mean) and their standard deviation. Higher values generally show greater dispersion. We may correct for this by expressing the standard deviation as a percentage of the mean using

$$CV = \frac{\sigma}{\bar{X}} \times 100 \tag{2.7}$$

For Kano annual rainfall (in cm), we have $\bar{X}_1 = 84 \cdot 66$, $\sigma_1 = 17 \cdot 35$ and

$$CV_1 = \frac{17 \cdot 35}{84 \cdot 66} \times 100 = 20 \cdot 5\%$$

and for Kano October rainfall (in mm), we obtain $\bar{X}_2 = 12 \cdot 4$, $\sigma_2 = 18 \cdot 7$ and

$$CV_2 = \frac{18 \cdot 7}{12 \cdot 4} \times 100 = 151\%$$

Although the standard deviation for October rainfall is minute ($\sigma^2 = 0 \cdot 187$ cm) compared with that for annual rainfall, it nevertheless represents a very much larger spread, relative to the mean. The coefficient of variation therefore gives a measure of *relative* variability.

> Is the coefficient of variation for the October rainfall figures changed if they are all expressed in centimetres rather than millimetres? Would it make any difference if they were expressed in inches?

EXERCISES I

1. The class structure of six Crawley New Town neighbourhoods in 1961 is shown below, with percentages in brackets.

	Area						
Class	West Green	Northgate	Three Bridges	Langley Green	Pound Hill	Gossops Green	Totals
I Professional	2(3·1)	2(2·1)	6(6·2)	2(1·2)	1(1·4)	8(8·1)	21
II Intermediate professional	9(14·1)	7(7·4)	9(9·3)	12(7·4)	13(18·6)	20(20·2)	70
III Unskilled non-manual; Skilled manual	38(59·4)	58(61·7)	64(66·0)	113(69·3)	44(62·9)	62(62·6)	379
IV Semi-skilled	11(17·2)	18(19·2)	15(15·4)	30(18·4)	7(10·0)	3(3·0)	84
V Unskilled manual	4(6·2)	9(9·6)	3(3·1)	6(3·7)	5(7·1)	6(6·1)	33
Totals	64	94	97	163	70	99	587

(a) Compare the neighbourhoods in terms of their social structure. Do you think social classes based entirely upon the occupation of the head of the household (in most cases taken to be the husband) are satisfactory?

(b) Calculate the percentages of middle-class and working-class households in each neighbourhood, taking classes I and II to represent the former, and classes III, IV and V to represent the latter.

(c) The concept of the 'socially balanced community' has often been discussed by sociologists and more recently by geographers. Taking the percentages above as given, in what sense do you think such communities might be created?

2. Take the Kano October rainfall observations, and calculate the variance and standard deviation using the grouped frequency method. Compare these with the correct values $\sigma^2 = 348\cdot7$ and $\sigma = 18\cdot7$. Why should there be such a discrepancy in this case, and what conclusions do you draw about the general applicability of the short-cut procedure?

3. Assemble all the measures of central tendency and dispersion discussed above with respect to the annual and October rainfall figures, and discuss their merits and limitations in terms of the contrasts they reveal.

4. Law and Warnes (1976) carried out a study of the distribution of the elderly, aged 60 or over, in 59 subdivisions of the standard regions of England and Wales, and also of the numbers of individuals aged 50–59. Summary statistics, all in percentages, are given below.

	Mean	Standard deviation	Maximum	Minimum
Elderly, 60 or over				
1951	16·65	2·34	24·08	12·72
1961	18·08	3·15	28·80	13·63
1971	19·97	3·77	32·61	14·84
Aged 50–59				
1961	13·21	0·86	15·38	11·21
1971	11·06	1·70	16·53	8·49

(a) Comment on any trends shown by the summary measures.
(b) What further summary measures should be calculated for purposes of comparison? Calculate them, and give a brief interpretation.
(c) According to Law and Warnes, the degree of skewness of the frequency distributions has become markedly more positive (i.e. right-skewed) over time. What has happened, therefore, to the percentage of elderly people in certain areas?

DESCRIPTIVE SPATIAL STATISTICS

A number of summary measures have been developed to describe spatial distributions (Bachi 1968, Neft 1966), and may be applied to the spatial distribution of phenomena represented as points, e.g. retail outlets, workplaces, residences, drumlins and inselbergs, or to administrative units, each of which is represented by a point such as the major town, or by a location judged to be close to the area's centre of gravity. For such **point patterns** we can define measures of central tendency and dispersion, exactly as was done earlier for non-spatial data.

Let us first consider **measures of central tendency**. The **median centre**, counterpart to the median, is located at the intersection of two axes placed at right angles to one another such that each axis has an equal number of points on either side. It is difficult to identify a unique median centre, because several pairs of axes, each intersecting at a different point, may fulfill this definition (Fig. 2.5). A computer routine is needed to determine the location of the 'true' median centre, which has the property of minimising the sum of the deviations of the observations about it, i.e.

$$\sum_{i=1}^{n} \mid (U_i, V_i) - M \mid = \text{a minimum}$$

where (U_i, V_i) represent the locational coordinates of the ith point, M is the median centre, and the two vertical lines mean that we ignore the sign of each deviation and effectively treat them as distances. One interpretation of the 'true' median centre is as the **point of minimum aggregate travel** (because of its distance minimisation property). Such a point may be of interest as it represents

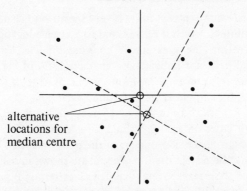

Figure 2.5   Locating the median centre.

the theoretical optimum location for a service facility such as a shop, school or warehouse, but Neft (1966, p. 31) comments 'that because of the difficulty in establishing its location, and its insensitivity to any change in the location of points within one of the quadrants defined by the orthogonal axes, the measure is 'virtually useless as a measure of average position'. However, insensitivity to items which show a highly eccentric location could be regarded as being among its virtues.

A more commonly employed measure of 'average position' is the **mean centre.** If the coordinates of any given point in a spatial distribution are (U_i, V_i) (Fig. 2.6), then the arithmetic mean centre is given by

$$(\bar{U}, \bar{V}) = \left(\frac{1}{n} \Sigma U_i, \frac{1}{n} \Sigma V_i \right) \tag{2.8}$$

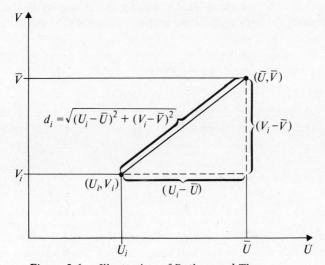

Figure 2.6   Illustration of Pythagoras' Theorem.

Like its non-spatial counterpart, $(\bar{U}, \bar{V})$ is very sensitive to changes in the spatial distribution of points to which it relates, and to 'extreme locations'. $(\bar{U}, \bar{V})$ need not be located within the area under consideration, as for example in an L-shaped area, and need not indicate any characteristics of the point where it is located – it represents an average *location*, not an average of the characteristics of the *phenomena* to be found at that location.

Henderson (1978) asked a number of 16-year olds in Reading and Walsall to locate 19 different cities on a map of England and Wales. Considering only the location of Birmingham, the coordinates of the points placed by the 41 children in Reading, and by the 51 children in Walsall, are shown in columns (1) and (2) of Tables 2.5a and 2.5b respectively. The true location of Birmingham is given by the coordinates $(41, 29)$, and the mean centre of the Reading group by

$$(\bar{U}, \bar{V}) = \left(\frac{1671}{41}, \frac{1483}{41}\right) = (40 \cdot 76, 36 \cdot 17)$$

Compared with the correct location, therefore, the average of those chosen by the Reading schoolchildren is biased northwards by 7 units, or approximately 35 km, since 1 unit = 5 km. It is interesting to speculate why there appears to be no overall east–west bias in the results.

A measure of **spatial dispersion** about the mean centre is the **standard distance deviation**, s_D. It is given by

$$s_D = \sqrt{(\Sigma d_i^2)/n} \tag{2.9}$$

where d_i^2 is the square of the distance between the location of the ith observation and that of the mean centre, $(\bar{U}, \bar{V})$. As shown in Figure 2.6, d_i is calculated using Pythagoras' theorem, so the definitional formula is

$$s_D = \sqrt{\frac{1}{n}(\Sigma[(U_i - \bar{U})^2 + (V_i - \bar{V})^2])} = \sqrt{\sigma_U^2 + \sigma_V^2} \tag{2.10}$$

showing that the standard distance is given by taking the square root of the sum of the variances of the northings (σ_U^2) and the eastings (σ_V^2). The computational formula is

$$s_D = \sqrt{\left(\frac{1}{n}\Sigma U_i^2 - \bar{U}^2\right) + \left(\frac{1}{n}\Sigma V_i^2 - \bar{V}^2\right)} \tag{2.11}$$

The deviation s_D may be computed with respect to the perceived locations of Birmingham chosen by the Reading schoolchildren, and taken as a measure of the agreement between them as to the city's location, irrespective of the degree of difference between the city's true location and $(\bar{U}, \bar{V})$. The smaller the value

Table 2.5   Coordinates of points representing Birmingham, as drawn on maps by schoolchildren in (a) Reading and (b) Walsall.

(a)

(b)

Easting	Northing			Easting	Northing
U_i	V_i	U_i^2	V_i^2	U_i	V_i
45	23	2025	529	35	35
42	33	1764	1089	41	27
42	42	1764	1764	43	27
43	26	1849	676	36	33
42	33	1764	1089	41	27
38	44	1444	1936	43	27
46	27	2116	729	36	35
43	33	1849	1089	41	28
40	45	1600	2025	44	23
49	26	2401	676	37	31
38	33	1444	1089	41	28
39	45	1521	2025	44	24
40	29	1600	841	38	27
41	34	1681	1156	41	32
37	46	1369	2116	44	26
42	29	1764	841	38	27
43	35	1849	1225	41	33
35	51	1225	2601	44	32
48	29	2304	841	38	30
32	36	1024	1296	41	35
33	53	1089	2809	45	25
49	29	2401	841	38	34
33	36	1089	1296	42	25
38	53	1444	2809	45	25
39	30	1521	900	39	25
43	37	1849	1369	42	26
28	58	784	3364	45	28
42	30	1764	900	39	26
53	35	2809	1225	42	27
41	31	1681	961	46	22
37	37	1369	1369	39	27
42	31	1764	961	42	27
43	37	1849	1369	46	25
44	31	1936	961	39	31
39	38	1521	1444	42	28
39	32	1521	1024	46	27
38	39	1444	1521	39	32
41	32	1681	1024	42	28
40	41	1600	1681	46	29
43	32	1849	1024	40	27
41	32	1681	1764	42	31
				46	31
Totals				41	24
1671	1483	69003	56249	42	31
				49	25
				41	25
				43	26
				50	22
				41	26
				43	27
				39	35
				2128	1434

$$\overline{U} = \frac{1}{n} \Sigma U_i = \frac{1671}{41} = 40{\cdot}76$$

$$\overline{V} = \frac{1}{n} \Sigma V_i = \frac{1483}{41} = 36{\cdot}17$$

of s_D, the greater the agreement. Using the data from Table 2.5a, we obtain

$$s_D = \sqrt{\left[\left(\frac{1}{41} \times 69003\right) - (40 \cdot 76)^2 + \left[\left(\frac{1}{41} \times 56249\right) - (36 \cdot 17)^2\right]\right.}$$

$$= \sqrt{85 \cdot 28} = 9 \cdot 23$$

This value of s_D may be plotted on a map by describing a circle of radius 9·23 about the mean centre (see Exercises II, below). Although useful, s_D does not tell us whether the spatial distribution departs from a circular pattern, as is clearly the case here since $\sigma_U^2 = 21 \cdot 62$ and $\sigma_V^2 = 63 \cdot 66$.

A better fit may be provided by the **standard ellipse**, which is also centred on the mean centre, and reflects certain directional biases in the data. However, the calculations are rather lengthy, and many of the formulae in standard texts are in error.

Although considered in the section on inferential statistics, the techniques for analysing point patterns (dot maps), described in Chapter 8, or area patterns (choropleth maps), described in Chapter 9, can also be used in a purely descriptive manner.

EXERCISES II

1. Use the data provided in Table 2.5 to determine the mean centre and standard distance with respect to locations for Birmingham chosen by the 51 Walsall children. You are also given that:

$$\Sigma U_i = 2128 \qquad \Sigma U_i^2 = 89306$$

$$\Sigma V_i = 1434 \qquad \Sigma V_i^2 = 40924$$

How do the results compare with those for the Reading children?

2. Plot the observations for both the Reading and Walsall children, using the grid system shown on pp. 8–9 of the *Ordnance Survey Gazetteer of Great Britain* (Ordnance Survey, 1972). Mark in the 'true' location of Birmingham and the two centres, and describe circles of appropriate radius about the mean centres to illustrate the standard distance concept.

REFERENCES AND FURTHER READING

Formal properties of scales of measurement are discussed by:
  Stevens, S. S. 1946. On the theory of scales of measurement. *Science* **103**, 677–80.
  Taylor, P. J. 1977. *Quantitative methods in geography*. Boston: Houghton Mifflin.
Concepts and operational definitions are treated in:
  Kerlinger, F. N. 1973. *Foundations of behavioural research*, 2nd edn. New York: Holt, Rinehart and Winston. (Ch. 3).

Nachmias, D. and C. Nachmias 1976. *Research methods in the social sciences.* London: Edward Arnold.

Descriptive statistics are widely used in population studies, as in:

Champion, A. G. 1976. Evolving patterns of population distribution in England and Wales 1951–71. *Trans Inst. Br. Geogs* (New Series) **1**(4), 401–20.

Goddard, J. B. and N. A. Spence 1976. *British cities – urban population and employment trends 1957–1971.* London: H.M.S.O., Dept of the Environment Research Report no. 10.

Law, C. M. and A. M. Warnes 1976. The changing geography of the elderly in England and Wales. *Trans Inst. Br. Geogs* (New Series) **1**(4), 453–71.

For an application in historical geography, see:

Langton, J. 1975. Residential patterns in pre-industrial cities: some case studies from seventeenth-century Britain. *Trans Inst. Br. Geogs* **65**, 1–28.

For a good example in urban studies, see:

Lewis, C. R. 1975. The analysis of changes in urban status: a case study in Mid-Wales and the middle Welsh borderland. *Trans Inst. Br. Geogs* **64**, 49–65.

The example based on accessibility to shops is from:

Guy, C. M. 1976. *The location of shops in the Reading area.* Reading: University of Reading, Dept of Geography, Geographical Paper no. 46.

The data on social class in Crawley New Town were taken from:

Heraud, B. J. 1968. Social class and the New Towns. *Urban Stud.* **5**(1), 33–58.

The class definitions were obtained from:

Classification of occupations, 1970 1970. London: H.M.S.O., Office of Population Censuses and Surveys.

Widespread use has been made of summary statistics by climatologists; see, for example:

Pye, N. 1972. Weather and climate. In *Leicester and its region* (Ch. 3). Leicester: Leicester University Press, for the British Association.

Smith, C. G. 1974. Monthly, seasonal and annual fluctuations of rainfall at Oxford since 1815. *Weather* **29**(1), 2–16.

The classic geomorphological studies carried out by Strahler and Schumm in the 1950s are referenced and summarised in:

Chorley, R. J. 1966. The application of statistical methods in geomorphology. In *Essays in geomorphology*, G. H. Dury (ed.). London: Heinemann.

For an example in soil studies, see:

Ball, D. F. and G. Williams 1968. Variability of soil chemical properties in two uncultivated brown earths. *J. Soil Sci.* **19**(2), 379–91.

Basic references on spatial descriptive statistics are:

Bachi, R. 1968. Statistical analysis of geographical series. In *Spatial analysis*, B. J. L. Berry and D. F. Marble (eds), 101–9. Englewood Cliffs: Prentice-Hall.

Neft, D. S. 1966. *Statistical analysis for areal distributions.* Philadelphia: Regional Science Research Institute, Monograph.

Applications to retail studies and consumer behaviour are:

Davies, R. L. 1972. The retail pattern of the central area of Coventry. In *The retail structure of cities*, 1–32. Institute of British Geographers, Occasional Paper no. 1.

Potter, R. B. 1977. The nature of consumer age fields in an urban environment: theoretical and empirical perspectives. *Tijdschr. Econ. Soc. Geog.* **68**(3), 168–76.

The techniques are well suited to the study of changes over time, and comparison of groups, as in:

Carter, F. W. 1975. C-K-D employees, Prague 1871–1920: some aspects of their geographical distribution. *J. Hist. Geog.* 1(1), 69–97.

Deskins, D. R. 1976. Race, residence and workplace in Detroit, 1880 to 1965. *Econ. Geog.* 52(1), 79–94.

The data on the children's perception were taken from:

Henderson, P. K. 1978. *The effect of spatial location on children's mental maps of Great Britain*. University of Reading, Dept of Geography, Unpublished undergraduate dissertation.

Other texts to which reference is made are:

Ebdon, D. 1977. *Statistics in geography: a practical approach*. Oxford: Basil Blackwell.

Norcliffe, G. B. 1977. *Inferential statistics for geographers*. London: Hutchinson.

3 Natural and Artificial Sampling Contexts

INTRODUCTION

Measurements must be viewed in a radically different manner if the concepts and techniques of inferential statistics, rather than of descriptive statistics, are to be used. Each measurement is held to consist of two components. The first component is the more important of the two, representing the influence of dominant factors which have a highly predictable influence upon the phenomena under investigation. Such influences are said to be **predictable** or **systematic**. The second component has an unpredictable influence on the value of any individual measurement, and represents the combined influence of a large number of less important factors. The combined influence of such factors is said to be **unsystematic** or **random** or **due to chance**.

There are two major contexts in which the geographer finds it useful to conceive of the existence of unpredictable or random elements in measurements.

NATURAL SAMPLING CONTEXTS

In natural sampling contexts, the 'natural' or 'socio-economic' system under investigation is thought of as producing phenomena, and hence measurements, which are in part the result of random or chance processes. For instance, earthquakes are generally associated with zones of weakness in the Earth's crust (systematic factor), but the time and place of their occurrence within those zones tends to vary from one year to the next (random or unsystematic factors). Similarly, the number of small firms moving each year from South-east England to development areas in England and Wales depends upon such factors as relative labour costs and accessibility to markets (systematic) and numerous minor factors such as the tastes and whims of the individual managing directors involved (random or unsystematic factors).

The notion that natural phenomena and social systems do not act in an entirely predictable manner is an attractive one for a number of reasons. First, it seems to accord with our experience. Secondly, it can be used to simplify the treatment of very complex phenomena by identifying a limited number of variables as systematic and representative of the most important processes or causes, and summarising the remainder as unsystematic or random. Indeed, this process of simplification, which owes much to the 'model building' movement in geography in the late 1960s, can almost be carried too far, as Curry has remarked that 'a probabilistic approach allows us to proceed on the basis of very little knowledge about the phenomenon under investigation'. Thirdly, it allows geographers to draw upon the powerful and elegant theory developed by statisticians, as does the use of artificial sampling procedures.

ARTIFICIAL SAMPLING CONTEXTS

In artificial sampling contexts, the investigator deliberately introduces a random or unsystematic element into the measurements obtained. This is done in the interests of efficient data collection where the total possible number of measurements of interest is extremely large or even infinite. A subset or **sample** of measurements is obtained by, in effect, selecting the items to be measured in a manner which includes a totally arbitrary or purely random element.

Because an unpredictable element is recognised (natural sampling context) or introduced (artificial sampling context), any conclusions based on such measurements must be stated in terms of probabilities. It is more realistic to talk of the probability of occurrence of a flood or an earthquake next year in a natural sampling context, or in an artificial sampling context of the probability that an estimate of the proportion of middle-class households in Langley Green, based on sample observations, lies within a specified margin of error from the true but unknown proportion. In the latter context, the true proportion of middle-class households may be regarded as the systematic or fixed component, and the difference between the true and estimated values as the unsystematic or random component.

Concern has been expressed in the geographic literature about the situations in which the concepts of inferential statistics should be applied. Meyer (1972a, b) argued that such concepts should be applicable only if measurements are obtained in an artificial sampling context, and King (1969) and Gould (1970) seem to agree with him. However, Court (1972) directly contests Meyer's views, and there is little doubt that geographers are now more likely to recognise the distinction between natural and artificial sampling contexts described above.

STOCHASTIC AND DETERMINISTIC PROCESSES

Any process which incorporates a truly random element is known as a **stochastic process**. As the preceding discussion suggests, the results of a stochastic process are not entirely predictable and may differ from one place or occasion to another, even if all the basic characteristics of the process remain unchanged. A **deterministic process**, on the other hand, always produces the same result if the basic characteristics of the process are unaltered. The difference between the two kinds of process will be illustrated clearly in Section II on natural sampling contexts.

The remainder of the book is divided into two major sections, the first devoted to natural sampling contexts, the second to artificial sampling contexts. Probability is a key concept considered in Chapter 4, and both this and the other two basically important notions of independence and randomness are introduced in various contexts in Chapters 4—9. Sampling procedures and a variety of methods for dealing with data obtained in an artificial sampling context are considered in Section III. It is important to recognise that many of the techniques discussed in Section III, although initially developed with an artificial sampling context in mind, can also be applied to measurements obtained in a natural sampling context. The interpretation of results must, however, be suitably modified.

REFERENCES AND FURTHER READING

Court, A. 1972. All statistical populations are estimated from samples. *Prof. Geog.* **24**(2), 160—1.

Gould, P. R. 1970. Is statistix inferens the geographical name for a wild goose? *Econ. Geog.* (Supplement) **46**(2), 439—50.

King, L. J. 1969. *Statistical analysis in geography*. Englewood Cliffs: Prentice-Hall. (pp. 60, 107, 123).

Meyer, D. R. 1972a. Geographical population data: statistical description not statistical inference. *Prof. Geog.* **24**(1), 26—8.

—— 1972b. Samples and populations: rejoinder to 'All statistical populations are estimated from samples'. *Prof. Geog.* **24**(2), 161—2.

Comments on the distinction between artificial and natural sampling contexts (although the writers do not employ these terms) may be found in:

Cliff, A. D. 1973. A note on statistical hypothesis testing. *Area* **5**(3), 240.

Haggett, P., A. D. Cliff and A. E. Frey 1977. *Locational analysis in human geography*, 2nd edn. Vol. 2: *Locational methods*. London: Edward Arnold. (p. 263).

Hammond, R. and P. S. McCullagh 1974. *Quantitative techniques in geography*. Oxford: Oxford University Press. (pp. 136—7).

An extended treatment is given in:

Gudgin, G. and J. B. Thornes 1974. Probability in geographic research: applications and problems. *The Statistician* **23**(3/4), 157—77.

Hepple, L. W. 1974. The impact of stochastic process theory upon spatial analysis in human geography. In *Progress in geography*, C. Board *et al* (eds). London: Edward Arnold.

II

NATURAL SAMPLING CONTEXTS

4　Probability

THE RELATIVE FREQUENCY DEFINITION OF PROBABILITY

The term 'probability' refers to the likelihood or chance of occurrence of a particular event. An **event** can be regarded as an experiment, as a point or period in time, as a point location or an area, or even as a household or individual, and must be capable of producing, or of being characterised by, at least two possible outcomes. Thus, a flume experiment can produce any of a variety of possible depths of scour within the limits set by an artificial channel, a period of time may be characterised by the outcomes 'flood' or 'non-flood', a point location by the presence or absence of a retail outlet, and a household may be characterised as having 0, 1, 2, 3, . . . children.

The event 'tossing a coin' has two possible outcomes, heads or tails, and it seems reasonable to believe that each outcome has the likelihood or chance of occurrence 0·5 or 50%. Perhaps what we do here is to imagine a very long sequence of events, in this case tosses of a coin. A practical way of checking our reasoning would be to note every occasion when a particular outcome *could* occur, and then record whether in fact it *has* occurred. A Danish statistician, Kerrich, found that of 10 000 coins tossed during his internment in World War 2, 5067 fell heads uppermost, so the proportion actually obtained (0·5067) seems fairly close to the 50% proportion we would expect. In most areas of interest to the geographer, it is not usually possible to specify probabilities *a priori* with any accuracy. To ascertain the likelihood of a 1 m or 2 m flood in a small river valley settlement in any given year, records must be examined over as long a period as possible for the settlement in question. In a 100-year record, for instance, it might be found that floods exceeding 1 m occurred in 13 years, while those exceeding 2 m occurred in four years. The proportions are thus 0·13 and 0·04 respectively.

The proportions observed in the coin-tossing and flood-occurrence examples are both based on the relative frequency of occurrence of events, and provide illustrations of a widely accepted definition of probability:

$$\text{probability of outcome} = \frac{\text{number of times outcome occurred}}{\text{number of events}} \quad (4.1)$$

where the number of events represents the maximum possible number of times the outcome *could* have occurred.

Our notion of 'probability' or 'likelihood' is the ratio of actual to potential occurrences based on as large a set of observations or experiments as possible.

The above formula represents the **relative frequency definition** of the probability of an outcome.

It is assumed that as the number of experiments performed or events observed increases indefinitely, so the proportion of times that a particular outcome occurs approaches a limiting value or **probability**.

> What is the limiting value in the case of the coin-tossing experiment? What do you think the value is with respect to the proportion of baby boys born in the United Kingdom?

A probability may simply be expressed as a ratio (e.g. 50/100 or 200/600), as a proportion (e.g. 0·5, 0·33), or as a percentage (e.g. 50%, 33%). It is most commonly written as a proportion or ratio, and will usually be expressed in this way in this text.

This relative frequency definition may seem reasonable, but it poses logical difficulties. Kerrich's coin-tossing experiment is very suggestive, but does not constitute proof that the probability of obtaining a head on any given toss is 0·5. Absolute verification really requires an experiment involving an infinite number of coin tosses, and this logical difficulty becomes a severely practical one if we are faced with estimating the probability of occurrence of processes — such as natural hazards — which are of high magnitude but low frequency. In many areas, reliable records may be available only for twenty or thirty years, or less, so that estimation of probabilities is virtually impossible. Even if we were sure about the relative frequency with which events occurred in the past, there is no logically necessary reason why this pattern should remain unchanged in the future. The number of cases required depends upon the kind of problem under consideration. In general, it is obviously impossible to demand an 'infinitely large' number of experiments.

It is worth noting that probabilities obtained by observation and experiment may not agree with those based on *a priori* and intuitive reasoning. The 'acid test' of our reasoning is empirical verification, bearing in mind the need for an appropriate number of cases.

STABILITY OF PROBABILITIES

It can be seen that probabilities may be difficult to estimate accurately. For instance, in ten tosses of a coin, seven may come up heads, and of ten children born in succession in the same hospital, only two may be girls.

The data in Table 4.1 show how proportions tend to stabilise as the number of events or length of a series increases. Ratios of male births to all births (sex ratios) vary between 0·512 and 0·518 for the major regions of England and for Wales, but between 0·488 and 0·613 for the rural districts of Dorset. The sex ratios are based on 50 000 births for most the major regions and about 300

Table 4.1   The sex ratio in England and Wales in 1973 – after an idea by Bulmer (1967, p. 3).

Regions of England and Wales	Sex ratio	Rural districts of Dorset	Sex ratio
North	0·518	Beaminster	0·543
Yorkshire and Humberside	0·516	Blandford	0·502
North-west	0·516	Bridport	0·488
East Midlands	0·514	Dorchester	0·552
West Midlands	0·516	Shaftesbury	0·504
East Anglia	0·512	Sherborne	0·505
South-east	0·516	Sturminster	0·613
South-west	0·518	Wareham and Purbeck	0·492
Wales	0·514	Wimborne and Cranborne	0·508
Both Countries	0·516	All RD's of Dorset	0·518

births for the rural districts. Greater stability seems to be associated with a larger sample size. This is also shown if the ratio for all rural districts, 0·518, is compared with that for both countries, 0·516. Notice also how an *a priori* probability of 0·5, which sounds 'reasonable', does not appear to be correct.

In the foregoing example, it is perhaps appropriate to assume that the probability in question is more or less constant over space and time, and that the larger the number of cases, the better the estimate. However, a different picture may emerge if the processes at work change over time. For example, it is known that the number of years in which any flood, defined as a day during which flow exceeded 4500 million gallons, occurred on the Thames at Teddington was 63 over the 80-year period 1883–1962 (Table 4.2). Over the same period, the number of years during which a 'moderate flood' (daily flow exceeding 7500 million gallons) occurred was 18. The long-run estimate of the probability of any given year experiencing a flood is therefore 63/80 or 0·788, and the corresponding figure for moderate floods 18/80 or 0·225. For each type of flood, the number of years in ten affected is noted for each consecutive period, as is the

Table 4.2   Ten-year and cumulative flood probabilities.

Ten-year sample period	All floods			Moderate floods		
	Number	Proportion	Cumulative proportion	Number	Proportion	Cumulative proportion
1883–1892	6	0·6	0·600	1	0·1	0·100
1893–1902	6	0·6	0·600	2	0·2	0·150
1903–12	8	0·8	0·667	2	0·2	0·167
1913–22	8	0·8	0·700	1	0·1	0·150
1923–32	8	0·8	0·720	4	0·4	0·200
1933–42	9	0·9	0·750	4	0·4	0·233
1943–52	9	0·9	0·771	2	0·2	0·229
1953–62	9	0·9	0·788	2	0·2	0·225

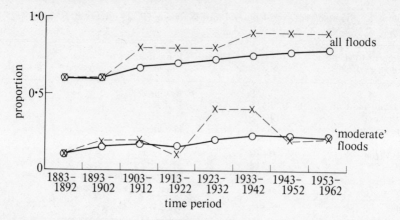

Figure 4.1   Short-run and cumulative proportions for Thames flood data (1883–1962). The broken lines and crosses show the proportion per 10-year period; the full lines and circles show the cumulative proportion.

cumulative proportion of floodyears. A plot of the sets of proportions shows how the proportions based on ten observations fluctuate quite widely about the long-run proportions, while the cumulative proportions also oscillate, but within a smaller range, and tend to vary much less as the number of observations upon which they are based increases (Fig. 4.1). The 10-year probabilities show an upward trend throughout the period. Although a long series of records is desirable when trying to predict 'catastrophic floods', some balance between the length of series and its relevance to current conditions must be achieved, particularly for more frequent events. In this case, the probability of a flood exceeding 4500 million gallons in one day in any given year is 0·788, based on an 80-year record, but that based on the 35-year period 1916–50 is 0·829. The latter figure is in fact based on a 'standard average period' defined by the Meteorological Office – a period thought long enough, and recent enough, to enable reliable estimates to be made for many purposes, and to permit comparison with figures for other areas and time periods.

In the case of the sex ratio, the proportions or probabilities apparently approach a common limiting value, whereas the same kind of pattern does not necessarily occur in the case of the flood data, or in many other situations of interest to the geographer.

PREDICTABILITY OF INDIVIDUAL EVENTS OR CASES

Although a probability of occurrence for a particular kind of outcome may be established, this does not mean we know, in any particular case, whether the outcome will occur or not. For instance, the probability of a flood-level river flow on the Thames in any one year was found to be approximately 0·80. Does this mean we can tell whether such a flow will occur on the Thames next year or not? The answer must be no, we cannot. It is more *likely* that there will be a flood than not, but we cannot be certain. The essence of a probabilistic statement is the element of uncertainty, and a willingness to state, for example, that on the basis of currently available information, the number of years in the next 100 that should experience flooding is 'close to' 80. A probability estimate allows us to make reasonably accurate predictions about the outcome of a large number or long series of events, not about any *single* event.

Lack of certainty about individual cases does not mean probabilities lack value. Suppose a new dam is to be constructed in a semi-arid region. It is important not to 'overdesign', as this involves extravagant expenditure, but nor should the dam be destroyed by a 'catastrophic event', e.g. a very high flood, during its economic lifetime of, say, 100 years. Suppose the choice is narrowed to two sites. At both, floods sufficient to jeopardise the project have occurred in the past, but records indicate that the probability of such an event in the next 105 years is 0·005 (or 1 in 200) at site A, and 0·001 (or 1 in 1000) at site B. If we choose site B, we might be unlucky and find that the 1 event in 1000 comes up within the next 105 years. However, we cannot know what will happen with certainty, and so come to a decision based on available information. Obviously, it is safer to choose site B than site A, where the likelihood of occurrence of this (still rare) outcome is five times as great.

SOME RULES OF PROBABILITY

A probability of 0·0 or 0% means that an event will never occur, whereas a probability of 1·0 means the event must occur. We cannot be surer than either of these two extremes, so that if $P(A)$ represents the probability of an outcome A, then

$$0·0 \leqslant P(A) \leqslant 1·0 \tag{4.2}$$

> Can you think of any phenomenon whose probability of occurrence is 0·0 or 1·0? If you find this difficult, what does it suggest about the nature of the 'real world' and of statistics?

If any two outcomes, A and B, of a given event are **mutually exclusive**, then the probability of either of the outcomes is equal to the sum of their individual probabilities. In symbols this is written as the **addition law**

$$P(A \text{ or } B) = P(A) + P(B) \tag{4.3}$$

So if A represents a move by a firm from London to South Wales, and B a move from London to Scotland, $P(A) = 0·012$ and $P(B) = 0·007$, the probability that a

Table 4.3   Results of hypothetical shopping survey.

Mode of transport	Shopping centre patronised		Total
	Suburbs	Elsewhere	
Car	150	650	800
Other	150	50	200
Total	300	700	1000

firm moves either to South Wales or to Scotland is $0·012 + 0·007 = 0·019$. When considering events that have only two possible outcomes, as in the case of floods (the outcomes being 'flood' or 'no flood'), then by Equation 4.2 P(flood) + P(no flood) = $1·0$. These rules apply also to events with three or more possible outcomes.

Assuming that the sample households referred to in Question 1, p. 21, are representative of all households, show that the probability
(i) that a West Green household is in social class I, II or III is 0·766, and
(ii) that any household in the six Crawley New Town neighbourhoods is working class is 0·845.

Greater care is required in calculating probabilities if the outcomes of an event are **non-mutually exclusive**. Suppose that 1000 households in a suburb of a large city were asked whether they usually did their main weekly grocery shopping at the local suburban shopping centre, or elsewhere, and also whether they made the trip by car or not (Table 4.3).

Applying the version of the addition law in Equation 4.3 the probability of a suburban shopping trip or a shopping trip by car, would be

$$P(\text{suburbs or car}) = P(\text{suburbs}) + P(\text{car})$$

$$= \frac{300}{1000} + \frac{800}{1000} = 1·1$$

which is obviously wrong because the probability exceeds $1·0$. The source of the error is that suburban trips and car trips are not mutually exclusive. Of the 300 trips to the suburbs, 150 are by car, so that these 150 trips are double-counted. To obtain the correct answer, we should subtract the probability of a 'suburbs *and* car' trip from the previous answer

$$P(\text{suburbs or car}) = P(\text{suburbs}) + P(\text{car}) - P(\text{suburbs and car})$$

$$= 0·30 + 0·80 - 0·15 = 0·95$$

The correction is obtained from the top left-hand corner of Table 4.3.

The general form of the **addition law** for non-mutually exclusive outcomes is

$$P(A \text{ or } B) = P(A) + P(B) - P(A \text{ and } B) \qquad (4.4)$$

> The correct answer to the 'suburbs or car' problem could have been obtained by applying the simpler version of the addition law to the four 'compound outcomes' in the body of Table 4.3. Show this to be the case.

EXERCISES

1. There has been rain on the day of Upper Dollop's agricultural fair, which is always held on the third Saturday in August, on 43 occasions in the last 100 years.
   (a) What is the probability of rain occurring on the day of the fair next year?
   (b) Will it rain on the day of the fair next year?
   (c) Is it more likely than not it will rain on the day of the fair next year?

2. Reliable records show a severe drought occurred in 15 of the last 100 years. Give a simple example to demonstrate that $p = 0.15$ is unlikely to give an accurate estimate of the probability of a severe drought next year.

3. The table below gives information on immigrant status ('New Commonwealth' or 'Other') and area of residence ('Inner London' or 'Outer London').

	A_1 Inner London	A_2 Outer London	Total
B_1 New Commonwealth	219 510	256 975	476 485
B_2 Other	299 405	293 010	592 415
Total	518 915	549 985	1068 900

Use the notation in the table (e.g. $P(A_1)$ = probability of living in Inner London: $P(A_1 \text{ or } B_2)$ = probability of living in Inner London or being an Other immigrant), when calculating the probabilities of
(a) being a New Commonwealth immigrant,
(b) living in Inner London,
(c) being a New Commonwealth immigrant and living in Inner London, and
(d) being a New Commonwealth Immigrant or living in Inner London. Show two ways of working out this probability.
Give the values of
(e) $P(A_1)$,
(f) $P(B_2 \text{ and } A_2)$, and
(g) $P(A_1 \text{ or } B_2)$.
   Do New Commonwealth immigrants seem more or less likely than Other immigrants to settle in Inner London? Does the answer surprise you in any way? How do you react if told the definitions of Inner and Outer London are

based on whole boroughs, as given in the GLC *Annual Abstract of Greater London Statistics 1971* (1972)?

REFERENCES AND FURTHER READING

An excellent discussion of probability concepts and probability laws is given in:
Bulmer, M. G. 1967. *Principles of statistics*, 2nd edn. Edinburgh: Oliver and Boyd. (Ch. 1).
In geography, study of probabilities has chiefly been the concern of climatologists, as in:
Court, A. 1953. Temperature extremes in the United States. *Geog. Rev.* **43**, 40—9.
Glover, J., P. Robinson and J. P. Henderson 1954. Provisional maps of the reliability of annual rainfall in East Africa. *Q. J. R. Meteorol. Soc.* **83**, 543—9.
Gregory, S. 1978. *Statistical methods and the geographer*, 4th edn. London: Longmans. (pp. 62—7). (Knowledge of the normal distribution is assumed.)
Rodda, J. C. 1967. A countrywide study of intense rainfall for the U.K. *J. Hydrol.* **5**, 58—69.
An application in medical geography is:
White, R. 1971. Probability maps of leukaemia mortalities in England and Wales. In *Readings in medical geography*, N. D. McGlashan (ed.). London: Methuen.
A probabilistic approach to regional analysis, based on an extension of the multiplication law, is given in:
Haggett, P., A. D. Cliff and A. E. Frey 1977. *Locational analysis in human geography*, 2nd edn. Vol. 2: *Locational methods*. London: Edward Arnold. (pp. 457—60).
The immigrant data are taken from:
Annual abstract of Greater London statistics 1971 1972. Vol. 6. London: Greater London Council.

5 Relationships Involving Characteristics or Attributes

INTRODUCTION

We can make use of the ideas introduced in Chapter 4 to analyse data if we think of persons, places or things as 'events', and any attribute or characteristic they possess as an 'outcome'. By an 'attribute' or 'characteristic' is meant a variable measured on a nominal scale. This includes a variety of socio-economic characteristics such as sex, race, social class and occupation and 'spatial' variables expressed in terms of regional categories such as 'the Midwest' or 'the South', or according to a set of administrative divisions for which published data are available. Many variables may be measured at the nominal rather than interval or ratio level in order to obtain some data rather than none at all—individuals are more willing to state that their ages or incomes fall within certain ranges than they are to give precise details. In other cases, statistical analysis may be possible only if measurements are converted to nominal form.

INDEPENDENCE OF CHARACTERISTICS

The statistical definition of independence is concerned with the relationship between two or more events. Suppose 100 suicides occurred in an inner city area of 100 000 people, and that the area consists of five districts with populations as shown in row (2) of Table 5.1a. We want to know if there is any tendency for suicides to be located in particular areas, and it seems reasonable to suppose that there is *no relationship* between district of residence and propensity to commit suicide if the number of suicides in each district is as shown in the first row of Table 5.1a. The probability of any given individual in the entire inner city area committing suicide is estimated by

$$P(S) = \frac{\text{total number of suicides}}{\text{total number of residents}} = \frac{100}{100\ 000} = 0 \cdot 001$$

Because $P(S)$ is calculated from the right-hand margin of Table 5.1a it is known as a **marginal probability**. If D_1 denotes 'resident in district 1', the probability of any given individual in district 1 committing suicide is estimated by

$$P(S \mid D_1) = \frac{\text{number of suicides in district 1}}{\text{number of residents in district 1}} = \frac{30}{30\ 000} = 0 \cdot 001$$

Table 5.1   Frequency distributions.

(a) Hypothetical frequency distributions of 100 suicides over five districts.

		\multicolumn{6}{c}{District}				
		1	2	3	4	5	Totals
	(1)	30	30	20	10	10	100
Population	(2)	30 000	30 000	20 000	10 000	10 000	100 000
Probability	(3)	0·001	?	?	?	?	0·001
	(4)	$P(S|D_1)$	$P(S|D_2)$	$P(S|D_3)$	$P(S|D_4)$	$P(S|D_5)$	$P(S)$
Numbers of	(5)	29	31	18	9	13	100
suicides	(6)	53	43	1	2	1	100
	(7)	35	33	13	4	15	100
	(8)	23	31	31	6	9	100

(b) Tally sheet for results of urn model experiment.

District	1	2	3	4	5
	11111 11111 11111 11111 11111 1	11111 11111 11111 11111 11111 11111	11111 11111 11111 11111 11	11111 11111 1111	11111 111
	26	30	22	14	8

(c) Frequency distribution averaged over 200 experiments.

	\multicolumn{5}{c}{District}				
	1	2	3	4	5
Totals generated in 200 runs	5888	6138	3992	1976	2006
Average over 200 runs	29·44	30·69	19·96	9·88	10·03

where $P(S|D_1)$ means 'the probability of an individual committing suicide given that he or she is resident in district 1'. The symbol (|) stands for 'given that' and $P(S|D_1)$ is called a **conditional probability**.

> Calculate the conditional probabilities with respect to the other four districts.

You should find that the conditional probabilities are equal to one another and to the marginal probability $P(S)$ (Table 5.1a). This is not surprising because it represents another way of stating that the propensity to commit suicide does not vary from district to district. In more formal language, the characteristics represented by the binary variable 'suicide or non-suicide' are statistically independent of the five categories of the nominal variable 'district of residence'.

The general definition of statistical independence states that

$$P(A \mid B) = P(A) \tag{5.1}$$

The probability of occurrence of event or characteristic A is therefore not influenced by the occurrence of event or characteristic B. The expression for the relationship between suicide and district of residence is of exactly the same form, but must include five conditional probabilities because the relationship involves a nominal variable with five categories.

If numbers of suicides in the inner city area were distributed over the five districts as shown in row (1) of Table 5.1a (30, 30, 20, 10, 10), or indicated only minor variations in the suicide rate as in row (5) (29, 31, 18, 9, 13), there would be little difficulty in concluding that suicide and district of residence were statistically independent (or virtually so), and no need to search for systematic factors to explain such variations; nor, at the other extreme, should there be any hesitation in searching for systematic factors to explain the high suicide rates in districts 1 and 2 as shown in row (6) of the table (53, 43, 1, 2, 1).

However, is it reasonable to presume that systematic factors are at work if the distributions (35, 33, 13, 4, 15) or (23, 31, 31, 6, 9) shown in rows (7) and (8) of Table 5.1a are observed? Statistical theory provides useful guidance in such cases and, to understand how, it is first necessary to introduce the concept of randomness.

RANDOMNESS

The concept of randomness is concerned with the way in which any individual outcome or characteristic occurs. Even if there were no relationship between suicide rates and district of residence, it does not seem likely that the total number of suicides in the inner city area would be spread perfectly evenly over the constituent districts. Slight variations from district to district are to be expected because of unpredictable or chance factors in the behaviour of individuals. The question is how big the variations have to be for us to believe that some factor or factors other than the minor ones summarised by chance are at work.

An answer may be provided by first assuming that suicide rates and district of residence *are* statistically independent. The total number of inner city suicides is then distributed over the five districts according to this assumption, *and* by making use of a chance mechanism. This may be done by placing 10 balls, three numbered 1, three numbered 2, two numbered 3, one numbered 4 and one numbered 5, in an urn. The numbers represent the districts. The urn is well shaken and a ball is then drawn out. Because each district is weighted according to its population, the probability of each *individual* being selected is the *same*, and the assumption of independence with respect to individuals fulfilled. Each ball is drawn quite arbitrarily, thus providing the random or chance element. After noting the number of the ball on a tally sheet, it is replaced, ensuring that any individual, no matter what her or his district of residence, has the same probability of selection on the second drawing. If the procedure is repeated 100 times, the result might be as shown in Table 5.1b. By chance alone, therefore, a frequency distribution has been produced which does not exactly conform to that ideally expected if there is no relationship between the two sets of characteristics. Repeating the entire process many times over, a slightly different frequency distribution might be obtained each time. Doing this 200 times in a class experiment, it was found that averaging over all 200 frequency distributions, the profile of the number of suicides closely approaches the limiting profile (30, 30, 20, 10, 10) (Table 5.1c).

The experiment provides an example of the use of what statisticians call an **urn model**. Because a ball is always replaced after each drawing, the procedure is referred to as **sampling with replacement**. Two important features of this experiment are as follows:

(a) In the long run, results should conform very closely to a set of limiting values. This confirms what has already been stated in the chapter on probability.

(b) Any particular set of results may differ to a varying degree from the ideal limiting values, and from other sets of results, but the *process* giving rise to them is the same in every case.

Both features are typical of a stochastic process. The process described above is an example of an **independent random** process, and the results differ from one 'run' to another, simply because of chance or random factors. The set of results obtained from each run of a stochastic process is known as a **realisation**.

> Suppose you were told that *exactly* 1 in 1000 (0·1%) of the residents of each of the five inner city districts will commit suicide each year. How many suicides would occur in each district? What kind of process is operating? How does it differ from a stochastic process?

THE CHI-SQUARE (χ^2) PROBABILITY DISTRIBUTION

We can take each set of experimental results, and calculate an index reflecting the overall deviation between them and the anticipated long-run frequencies. This index is called **chi-square** (chi rhyming with sky) and denoted χ^2. If the observed frequency in the ith district (or category) is represented by o_i, and the corresponding long-run frequency by e_i, then

$$\chi^2_\nu = \sum_{i=1}^{k} \left(\frac{(o_i - e_i)^2}{e_i} \right) \qquad (5.2)$$

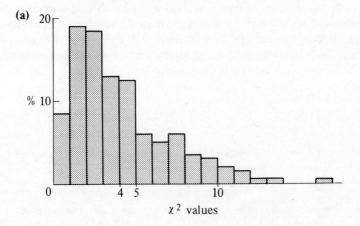

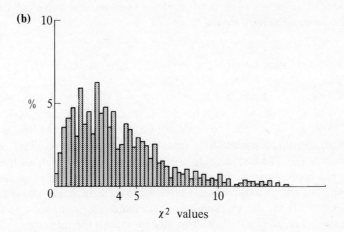

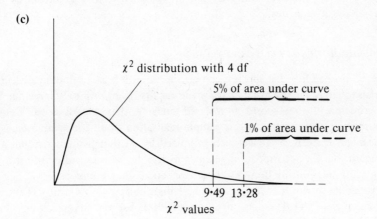

Figure 5.1   Experimentally and theoretically derived χ^2 distributions based on four degrees of freedom.

where k is the number of categories, and ν the number of **degrees of freedom** (df) associated with the χ^2 value and given by the number of categories minus one, i.e. $(k-1)$. So df $= (5-1) = 4$ here. The meaning of this term will be explained fully later. Applied to the first set of observed frequencies obtained from our experiment (Table 5.1c), we have

$$\chi_4^2 = \frac{(26-30)^2}{30} + \frac{(30-30)^2}{30} + \frac{(22-20)^2}{20} + \frac{(14-10)^2}{10} + \frac{(8-10)^2}{10} = 2\cdot733$$

Notice that the wider the discrepancy between observed and expected frequencies, the greater will be the value of χ^2.

> Check this by showing that the χ^2 value for the observed frequencies (53, 43, 1, 2, 1) in row (6) of Table 5.1a is 55·816. What is the value of χ^2 if the observed and expected frequency distributions are identical?

The calculation may be repeated with respect to all 200 sets of frequency distributions produced by the experiment, and the χ^2 values plotted in a histogram (Fig. 5.1), giving a picture of the overall form of the frequency distribution of χ^2 values with 4 df. If the results of, say, 1000 experiments were available, it would be possible to narrow the class interval of the histogram — results obtained using a computer to imitate our urn model procedure are also given in Figure 5.1. The results of an indefinitely large number of such experiments should approximate a limiting distribution known as the χ^2 **probability distribution**. The continuous curve in Figure 5.1 is mathematically derived under this assumption. Thus, the χ^2 probability distribution is continuous, in theory taking on any value from zero to plus infinity $(+\infty)$, and the total area under the curve represents a probability of $1\cdot0$ or 100%. Note that χ^2 obviously cannot take on negative values.

Sampling distribution and expected value

The χ^2 probability distribution can also be regarded as a **sampling distribution**, because it may be constructed from the results of an infinitely large number of experiments. In real-world terms, we imagine that the observed frequency distribution represents just one **sample realisation** of all the sample realisations that might conceivably have been produced by an independent random process representing the occurrence of suicides in five districts according to the assumption of no relationship between these two sets of characteristics.

It can be shown that the mean of a χ^2 distribution with 4 df is 4 (i.e. equal to the number of df). The mean based on the 200 observed frequency distributions is $4\cdot083$, and that based on 1000 is $4\cdot060$, and most of the values in Figure 5.1 are grouped fairly close to this value. The limiting mean value towards which the χ^2

values tend is called the **expected value** and is the term reserved by statisticians for the mean of a probability distribution. Strictly speaking, therefore, whenever we talk of the 'mean' of a probability distribution we are referring to its expected value.

Looking up critical values in χ^2 tables

Large deviations between observed and expected values, as measured by the χ^2 index, *can* occur, but from Figure 5.1 we see that these are rare and are therefore *unlikely* outcomes if our expected frequencies are a correct representation of the underlying process. The values in the body of Table A.1 (see Appendix) are known as **critical points** or **percentage points** of the theoretically derived χ^2 distribution. This is because a specified percentage of the area under the curve is to be found *above* each point. Let us make use of Table A.1 with respect to the χ^2 distribution with 4 df. The entry at the intersection of v(or df) = 4 and column $P = 0.05$ tells us the probability of obtaining a χ_4^2 value greater than or equal to 9·49 is 0·05 or 5%, i.e. $P(\chi_4^2 \geqslant 9.49) = 0.05$, and 0·05 or 5% of the area under the curve lies to the right of $\chi^2 = 9.49$ in Figure 5.1c. The entry at the intersection of df = 4 and $P = 0.01$ is 13·28 so that $P(\chi_4^2 \geqslant 13.28) = 0.01$.

Experimentally derived class results agree quite closely with the theoretical results. In 200 experiments, the number of χ_4^2 values exceeding 9·49 and 13·28 were 12 (6%) and 3 (1.5%) – in 1000 experiments, the corresponding values were 44 (4·4%) and 8 (0·8%).

Table A.1 may now be used to find how likely it is that any observed frequency distribution will occur, assuming no relationship between (or statistical independence of) suicide and district of residence.

Intuitively, we expect the frequency distribution (29, 31, 18, 9, 13) in row (5) of Table 5.1a to be among the 'least extreme' liable to occur. The value of χ^2 is

$$\chi_4^2 = \frac{(29-30)^2}{30} + \frac{(31-30)^2}{30} + \frac{(18-20)^2}{20} + \frac{(9-10)^2}{10} + \frac{(13-10)^2}{10} = 1.266$$

The entry at the intersection of df = 4 and $P = 0.90$ is 1·06, so that $P(\chi_4^2 \geqslant 1.06) = 0.90$. Thus, the observed value of 1·266 cannot be regarded as very extreme because it occurs within the region under the probability curve where 90% of all the values of χ_4^2 obtained purely by chance would be expected if there were no relationship between suicide and district of residence. A reasonable person might well conclude that there is insufficient evidence to reject the hypothesis of no relationship.

The frequency distribution (53, 43, 1, 2, 1) in row (6) of Table 5.1a looks far more extreme, as already shown by the associated χ_4^2 value of 55·816. From Table A.1 we find that the probability of obtaining a value greater than or equal to 14·86 is 0·005 (0·5% or 1 in 200). Because it falls above 14·86, the value

55.816 cannot occur with a probability greater than 0.001 or 0.1%. As this probability of obtaining such a value, assuming no relationship between suicide and district of residence, is so small, there is every reason to doubt that our assumption was correct. A reasonable person might well believe that the difference between the anticipated and observed frequency distributions does not simply reflect chance factors, but also a systematic difference in suicide rates between districts.

The probabilistic argument outlined above typifies the logic underlying a statistical test of significance.

Use Table A.1 to find the *probability* that χ_4^2 (i) exceeds 0.43, (ii) exceeds 7.78, (iii) falls between 0.30 and 13.28, and the *value* of χ_4^2 that is exceeded with probability (iv) 50%, (v) 0.1%, (vi) 30%.

THE χ^2 TEST

A statistical test of significance based on the properties of the χ^2 distribution is known as a χ^2 **test**. The test may be described in a series of steps.

(1) Set up two **statistical hypotheses**, the **null hypothesis** H_0 and the **alternative hypothesis** H_1. In general, the null hypothesis represents a 'straw man' the researcher hopes to knock down so that the alternative hypothesis may be accepted. For the suicide and district of residence problem, the statistical hypotheses are
H_0: there is no relationship between suicide rates and district of residence,
H_1: there is a relationship between suicide rates and district of residence.
In symbols, this may be written
$H_0: P_1 = P_2 = P_3 = P_4 = P_5 = P,$
H_1: at least two of the P_i are unequal, where $P_1 = P(S|D_1)$, $P_2 = P(S|D_2)$ etc.

(2) Determine the expected frequency distribution of suicides 'under H_0', that is, if the null hypothesis is assumed to be true. The expected frequency of deaths in each district will be $P \times$ (population of district). However, the true value of P is unknown, because the observed number of suicides, as well as their frequency distribution over the five districts, could well vary from one realisation of an independent random process to another, and so is estimated from the observations. This estimate is denoted by p. If o_i denotes the observed number of suicides in the ith district, and n_i the population of the ith district, then we have

$$p = \frac{\Sigma o_i}{\Sigma n_i} \qquad (5.3)$$

so in this case $p = 100/100\ 000 = 0\cdot001$. In order to estimate e_i, we calculate

$$e_i = p \times n_i \qquad\qquad (5.4)$$

Applied to the first district, this gives $e_1 = 0\cdot001 \times 30\ 000 = 30$. Formula 5.4 may be expanded if we substitute from Formula 5.3 and then rearrange so that

$$e_i = p \times n_i = \frac{\Sigma o_i}{\Sigma n_i} \times n_i = \frac{n_i}{\Sigma n_i} \times \Sigma o_i \qquad\qquad (5.5)$$

showing that the total number of suicides is distributed over the districts according to their relative population sizes.

> Check that Formulae 5.4 and 5.5 give the same expected frequencies for the five districts, and that the sum of the expected frequencies is equal to the sum of the observed frequencies.

(3) Calculate the value of χ^2 which, when used to carry out a test of significance, is called the **test statistic**. Earlier, we wanted to decide whether the observed frequency distribution (35, 33, 13, 4, 15) was likely to have been produced by a purely independent random process, indicating no relationship, or not. Calculation of the test statistic shows $\chi^2_4 = 9\cdot683$.

(4) Determine the number of degrees of freedom v. The *maximum* possible number of degrees of freedom here is given by the number of categories, $k = 5$, because each observation (suicide) is 'free' to locate itself in any one of five categories. In general, v is obtained by subtracting 1 df from the total number possible for each linear constraint placed on the observations. Only one such constraint has been imposed here, namely

$$e_1 + e_2 + e_3 + e_4 + e_5 = 100$$

i.e. the sum of the expected frequencies must equal 100. Thus, $v = (k - 1) = 4$.

(5) Decide how 'extreme', in probabilistic terms, the value of χ^2 has to be in order for us to believe that H_0 is false. Conventionally, if the value of the χ^2 test statistic is likely to occur not more than 5% of the time under H_0, then H_0 is rejected. The probability level at which we decide to reject H_0 is called the **significance level** of a statistical test, and is denoted by α. Thus the significance level is $\alpha = 0\cdot05$ or $\alpha \times 100 = 5\%$ in this instance. (Note that α is always expressed as a decimal proportion, but significance levels are often quoted in percentage terms. Both forms are used in this text.) The critical point or percentage point corresponding to $\alpha = 0\cdot05$ is the value of χ^2_4 equalled or exceeded with a probability of $0\cdot05$ or 5% when H_0 is true. This critical value may be read from Table A.1 and is $9\cdot49$.

(6) Decide whether or not to reject H_0. The shaded area lying to the right of $\chi_4^2 = 9\cdot49$ in Figure 5.1 is called the **region of rejection**. If the value of the test statistic calculated from our data falls within this region, H_0 may be rejected at the $\alpha = 0\cdot05$ or 5% significance level. Since $\chi_4^2 = 9\cdot683$, we are just able to reject H_0 at this level of significance. To be more certain that the observed and expected frequencies are really generated by different processes, set $\alpha = 0\cdot01$ (or 1%). The region of rejection now lies to the right of $\chi_4^2 = 13\cdot28$ (Fig. 5.1). Our calculated value of $9\cdot683$ does not fall in the region of rejection defining the most extreme 1% of values obtained under H_0, so H_0 cannot be rejected at this level of significance.

Type 1 and type 2 errors

It cannot be emphasised too strongly that the conclusion to which we come, based on a statistical test of significance, is probabilistic. There is no way we can be *certain* that H_0 is true or false. If we decide to reject H_0, there is a chance that we are mistaken. The maximum probability of rejecting H_0 when it is in fact true is given by the significance level α, and the error made in rejecting H_0 when it is true is called a **Type 1 error**. If we had set $\alpha = 0\cdot10$ (10%) then a less extreme value of χ_4^2 would enable us to reject H_0, but the risk of a type 1 error is increased. To reduce the risk of such an error, we might set $\alpha = 0\cdot01$ (1%). This is the smallest value of α that is employed under normal circumstances. Any further reduction in α, and in the risk of committing a type 1 error, is counterbalanced by the increasing risk of committing a **Type 2 error** — that of failing to reject H_0 when it is in fact false. The relationship between our statistical conclusion and the truth or falsity of H_0 is summarised in Table 5.2. In general, the smaller the value of α, the more difficult it is for the test to distinguish between values of χ^2 generated by the difference between a set of observations and those expected under the null hypothesis of independence (30, 30, 20, 10, 10) and those expected under some other null hypothesis which is not too markedly different, such as (28, 32, 22, 6, 12), but which does not conform to the assumption of independence. Provided α is kept between $0\cdot05$ and $0\cdot01$, a reasonable compromise is achieved between the risks of committing a Type 1 or Type 2 error.

Table 5.2   Relationship between statistical conclusion and status of H_0.

Conclusion	Status of null hypothesis	
	H_0 true	H_0 false
H_0 not rejected	✓	Type 2 error
H_0 rejected	Type 1 error	✓

EXAMPLE 53

One- and two-tailed tests

The version of the χ^2 test described here is a **one-tailed** test, because we are interested only in values located in one 'tail' of the χ^2 distribution, in this case the extreme right-hand or *upper* tail (Fig. 5.1). Examples of tests involving lower tails of probability distributions, or both tails (**two-tailed** tests) will be discussed as they arise in subsequent chapters.

Assumptions underlying use of the χ^2 test

The χ^2 test should be used only if the following assumptions are fulfilled:

(1) The values of e_i and o_i are in the form of frequencies or counts obtained in a number of categories. Percentages, proportions or rates per thousand etc must not be used, unless a special version of the test is employed (Maxwell 1961, Ch. 6, Lewis 1977, pp. 128–47).
(2) The sum of the frequencies must be greater than 20 and, preferably, greater than 40.
(3) In any one category, the *expected* frequency should not normally be less than 5. However, if there are five or more categories then not more than 20% of the expected frequencies may be less than 5, and there should be no category with an expected frequency less than 1.
(4) Whatever systematic variations exist in the observations, there should also be a component which may be regarded as independent and random.

Example

The number of suicides reported in each of the metropolitan counties of England, and the population 'at risk', i.e. aged 15 or over, are shown in Table 5.3. Population figures are correct to the nearest hundred. Is there any relationship between suicide rate and metropolitan region of residence?

> Set up the appropriate statistical hypothesis and select a significance level.

Using Formula 5.5, and substituting the appropriate values from Table 5.3, the expected frequency in Greater London is

$$e_1 = \frac{5\,678\,400}{14\,524\,500} \times 1578 = 616 \cdot 924$$

and in Greater Manchester is

$$e_2 = \frac{2\ 060\ 300}{14\ 524\ 500} \times 1578 = 223 \cdot 839$$

> Using a hand calculator, obtain the remaining expected frequencies, and enter them in Table 5.3. Show that the value of χ^2 is $55 \cdot 913$. Decide whether there is any relationship between suicide rates and the nominal variable 'metropolitan region'.

Although the test itself must be based on frequencies, the results may be discussed in terms of probabilities, proportions or rates.

It should be remembered that different results may be obtained at different levels of temporal and spatial aggregation. If suicides for the same regions over a five-year period are examined, for instance, would a similar pattern emerge? Greater reliance may be placed upon results obtained from a longer period, and thus from a larger number of observations. Similarly, a breakdown of the Greater London figures into those for individual boroughs, or for Inner and Outer boroughs, may show variation within the region (see Q. 1, p. 62).

The χ^2 test may be regarded both as a test of **goodness of fit**, i.e. how well the observed frequency distribution fits a hypothetical expected frequency distribution, or as a test of **association**, i.e. for the existence of a relationship between nominal variables. So far, we have emphasised the former role of χ^2.

Table 5.3   Number of suicides and population by metropolitan region in England (1974).

Region	Observed suicides o_i	Suicide rates (per 1000)	Population	Expected suicides e_i
Greater London	749	0·1319	5678 400	616·924
Greater Manchester	213	0·1034	2060 300	223·839
Merseyside	116	0·0967	1200 100	? 130.389
South Yorkshire	78	0·0780	999 500	? 108.590
Tyne and Wear	100	0·1102	907 500	? 98.594
West Midland	174	0·0831	2093 800	? 227.479
West Yorkshire	148	0·0934	1584 900	? 172.190
Totals	1578	0·1086 (Average)	14 524 500	1578·0

χ^2 AS A TEST OF ASSOCIATION

Two characteristics of a person, place or other event may or may not tend to go together. Suppose a small area of downland is divided up into 36 square cells or

(a) Perfect positive association

A+B	A+B	A+B	A+B	A+B	A+B
A+B	A+B	A+B	A+B	A+B	A+B
A+B	A+B	A+B	A+B	A+B	A+B
A+B	A+B	A+B	A+B	A+B	
A+B	A+B	A+B	A+B		
A+B	A+B	A+B			

(b) Perfect negative association

A	A	A	A	A	B
A	A	A	A	B	B
A	A	A	B	B	B
A	A	B	B	B	B
A	B	B	B	B	B
B	B	B	B	B	B

(c) Tendency to positive association

A+B	A+B	A+B		A+B	A+B
A+B	B	A	B		
A+B	A+B	A+B	B	A+B	A+B
B	A	A+B	A+B	A+B	
	A+B		A+B	B	
A+B	A+B	B	A+B		A+B

(d) Tendency to negative association

A	A	A	A	B	
A	A+B	A	B	A	B
A	A	B	B	A	A
A	A	A+B	B	B	B
B	A+B	A	B	A+B	B
B	A	A	B	B	B

Figure 5.2 Relationships in space between two plant species.

quadrats, and the presence (+) or absence (−) of two different plant species, A and B, is recorded for each quadrat. There are four possible outcomes: (i) both species present, (ii) only species A present, (iii) only species B present, (iv) both species absent. If the two species are always found in the same quadrat, but never alone, then the pattern of results shown in Figure 5.2a might result, and counting of outcomes yields the **contingency table** of Table 5.4a. This is called a 2 × 2 (two by two) contingency table as each characteristic has two classes. The pattern represents one of perfect **positive association**, since the two species always go together. On the other hand, if the two species are never found together in the same quadrat, the patterns of Figure 5.2b and Table 5.4b might result, showing perfect **negative association**. Both relationships are deterministic, and it is more usual to find a *tendency* towards positive association (Fig. 5.2c and Table 5.4c) or negative association (Fig. 5.2d).

> Fill in the blanks in Table 5.4d by counting outcomes in Figure 5.2d.

Table 5.4   Relationships between two plant species (A and B) classified according to presence (+) or absence (−).

(a) Perfect positive association.

	B+	B−	Total
A+	30	0	30
A−	0	6	6
Total	30	6	36

(b) Perfect negative association.

	B+	B−	Total
A+	0	15	15
A−	21	0	21
Total	21	15	36

(c) Tendency to positive association.

	B+	B−	Total
A+	20	2	22
A−	6	8	14
Total	26	10	36

(d) Tendency to negative association.

	B+	B−	Total
A+	?	?	20
A−	?	?	16
Total	19	17	36

(e) Expected values based on marginal totals in (d).

	B+	B−	Total
A+	10·556	9·444	20
A−	8·444	7·556	16
Total	19	17	36

(f) Observations generated by urn model.

	B+	B−	Total
A+	12	?	20
A−	?	?	16
Total	19	17	36

Assuming no association, positive or negative, between the two plant species, what kind of pattern is to be expected? One possibility is given in Table 5.4e which you should notice has the same marginal (i.e. row and column) totals as Table 5.4d. (The fact that the expected frequencies need not be whole numbers will be explained shortly.) From the first column of Table 5.4e, the probability that plants of species A will be associated with plants of species B is

$$P(A+ \mid B+) = \frac{10 \cdot 556}{19} = 0 \cdot 556$$

where $P(A+ \mid B+)$ is the conditional probability that species A is present given that species B is present. Similarly, the probability that species A is present, given that species B is absent, is

$$P(A+ \mid B-) = \frac{9 \cdot 444}{17} = 0 \cdot 556$$

From the marginal totals in Table 5.4e, we obtain

$$P(A+) = \frac{20}{36} = 0 \cdot 556$$

and so

$$P(A+ \mid B+) = P(A+ \mid B-) = P(A+)$$

fulfilling the statistical definition of independence.

> Show also that $P(A- \mid B+)$, $P(A- \mid B-)$ and $P(A-)$ are equal.

.444   .444   .444

Thus, no association between characteristics implies statistical independence, and the information in Table 5.4e may be used, firstly, to demonstrate an important law governing the relationship between probabilities of independent events and, secondly, to show how expected frequencies under the null hypothesis of no association may be derived using this law.

The multiplication law for independent events

If A and B are two possible outcomes of events which are statistically independent, then

$$P(A \text{ and } B) = P(A) \times P(B) \qquad\qquad (5.6)$$

The probability that a quadrat referred to in Table 5.4e possesses characteristic A+ *and* characteristic B+ should therefore be equal to the product of the marginal probabilities $P(A+)$ and $P(B+)$. We find that

$$P(A+ \text{ and } B+) = \frac{10 \cdot 556}{36} = 0 \cdot 293$$

and

$$P(A+) \times P(B+) = \frac{20}{36} \times \frac{19}{36} = 0 \cdot 293$$

as anticipated.

> Show that the same relationship holds for $P(A+ \text{ and } B-)$, $P(B+ \text{ and } A-)$ and $P(B- \text{ and } A-)$ for the appropriate marginal probabilities in Table 5.4e.

If the row totals of Table 5.4e are denoted r_1 and r_2, the column totals by c_1 and c_2, and the grand total by T, then the probabilities calculated above may be written

$$P(A+ \text{ and } B+) = \frac{r_1}{T} \times \frac{c_1}{T} = \frac{r_1 c_1}{T^2}$$

So in general the expected frequency in the cell in the ith row and jth column is obtained if we multiply T by the appropriate marginal probabilities, i.e.

$$e_{i,j} = \frac{r_i c_j}{T^2} \times T = \frac{r_i c_j}{T} \tag{5.7}$$

Applied to the cell in the 1st row and 1st column of Table 5.4e, this gives

$$e_{1,1} = \frac{20 \times 19}{36} = 10 \cdot 556$$

Show that $e_{1,2}$, $e_{2,1}$ and $e_{2,2}$ equal $9 \cdot 444$, $8 \cdot 444$ and $7 \cdot 556$ respectively in Table 5.4e.

This set of expected frequencies should be approximated if, holding the *marginal* and *overall* totals constant, we assign outcomes randomly to cells over a large number of experiments, and then calculate the average number of outcomes per cell. Notice that this explains why the expected frequencies need not be whole numbers.

One way of carrying out the experiment is to select just *one* cell in the contingency table, for example that representing quadrats with both species (A+ and B+), and make use of the property that $P(A+|B+) = P(A+) = 0 \cdot 556$ if the characteristics are independent. Here, this means each quadrat containing plants of species B has a $0 \cdot 556$ (55·6%) chance of also containing plants of species A. There are 19 quadrats containing plants of species B, so in terms of an urn model we place 1000 balls, numbered 1 to 1000, in an urn and make 19 drawings with replacement. If any number in the range 1 to 556 inclusive is drawn, the quadrat also contains plants of species A and a tally mark is recorded in cell (A+, B+) (Table 5.4f); otherwise, nothing is recorded. Suppose as a result there are 12 entries in cell (A+, B+). We can immediately 'fill in the blanks' in Table 5.4f.

Fill in the blanks in Table 5.4f.

The set of outcomes in Table 5.4f may be regarded as one possible realisation of an independent random process which allocates outcomes to one cell under the assumption of independence of the characteristics represented by the rows and columns of the table. Numerous repetitions followed by averaging should yield cell values quite close to those in Table 5.4e.

Notice that contingency tables with *different* sets of marginal totals will yield *different* sets of expected values.

Comparison of a set of observed frequencies with a set of long-run expected

EXAMPLE 59

frequencies may be based on the test statistic

$$\chi_v^2 = \sum_{i=1}^{R} \sum_{j=1}^{C} \left(\frac{(o_{ij} - e_{ij})^2}{e_{ij}} \right) \qquad (5.8)$$

where o_{ij} and e_{ij} are the observed and expected frequencies in the cell in the ith row and jth column respectively, and the double summation $(\Sigma\Sigma)$ instructs us to sum over the R rows and C columns of the table, working along each row in turn. For the o_{ij} and e_{ij} in Tables 5.4e and 5.4f, we have

$$\chi_v^2 = \frac{(o_{11} - e_{11})^2}{e_{11}} + \frac{(o_{12} - e_{12})^2}{e_{12}} + \frac{(o_{21} - e_{21})^2}{e_{21}} + \frac{(o_{22} - e_{22})^2}{e_{22}}$$

$$= \frac{(12 - 10.556)^2}{10.556} + \frac{(8 - 9.444)^2}{9.444} + \frac{(7 - 8.444)^2}{8.444} + \frac{(9 - 7.556)^2}{7.556} = 0.942$$

The degrees of freedom are given by (number of rows minus one) x (number of columns minus one), i.e. $(R - 1) \times (C - 1)$. As there are two rows and columns here, $R = 2$, $C = 2$ and $v = (2 - 1) \times (2 - 1) = 1$. The maximum value for df is given by $(R \times C) = 4$ (there are four cells into which an outcome might fall), and knowledge of a value for any one cell of the contingency table automatically determines those of the other three — as shown by the workings of the urn model. Thus, we are left with only $(4 - 3) = 1$ df for the χ^2 test statistic.

The difference between observed and expected frequencies in a contingency table based on two nominal variables, each with two categories, as measured by χ^2, approximately follows a χ^2 distribution with 1 degree of freedom.

Use Table A.1 to show that the χ^2 value of 0.942, based on 1 df, is not, in probabilistic terms, an 'extreme' outcome.

Example

The spatial distribution of two plants, the wood sage (*Teucrium scorodonia*) and the foxglove (*Digitalis purpurea*) in Hertfordshire, England, was plotted in 446 quadrats covering the county by Doney (Seddon 1971, p. 63) (Fig. 5.3).

Before testing for association, set up the hypotheses

H_0: there is no association between the locations of wood sage and foxglove,
H_1: there is an association between the locations of wood sage and foxglove.

Observed and expected frequencies are given in Table 5.5, df = $(2 - 1) \times (2 - 1) = 1$ and $\chi_1^2 = 67.012$.

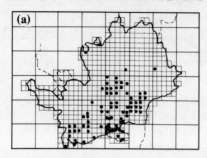 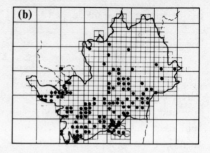

Figure 5.3   Spatial distribution of (a) wood sage and (b) foxglove in Hertfordshire.

> Use Table A.1 to show that the association between the two species is significant at both the 5% and 1% significance levels. Is the association positive or negative?

If the total number of observations in a 2 x 2 contingency table is relatively low – say, less than 40 – a better approximation to the χ^2 distribution with 1 df is given by

$$\chi_1^2 = \Sigma\Sigma \frac{(|o_{ij} - e_{ij}| - 0\cdot5)^2}{e_{ij}} \tag{5.9}$$

Here the sign of the difference between each pair of o_{ij} and e_{ij} is ignored, as indicated by the absolute value symbol $|\ldots|$, and we then subtract $0\cdot5$ from each difference before squaring.

Associations may also be revealed by the investigation of more complex contingency tables. For example, Cohen and Berry (1975) studied the relationship between manufacturing change (measured as change in the total number of manufacturing establishments) and size of city for 3102 US counties between 1952 and 1962. In Table 5.6, A_1 represents 'small cities' (core population under 25 000), A_2 'medium cities' (core population 25 000 – 100 000) and A_3 'large cities' (core population exceeding 100 000). The table indicates a positive

Table 5.5   Frequencies of occurrence of wood sage (W) and foxglove (F) in Hertfordshire (+ present; – absent).

(a) Observed				(b) Expected			
	W+	W–	Total		W+	W–	Total
F+	50	77	127	F+	21	106	127
F–	24	295	319	F–	53	266	319
Total	74	372	446	Total	74	372	446

Table 5.6   Association between manufacturing change (1952—62) and place in the urban hierarchy (1952).

| Manufacturing change | Place in hierarchy[*] | | | Total |
	A_1 (small)	A_2 (medium)	A_3 (large)	
Decline	[397	263	209]	869
Growth	[671	756	806]	2233
Total	1068	1019	1015	3102

association between core city size and manufacturing growth, and comparison of the observed and expected frequencies (in brackets) gives a value of $\chi^2_2 = 75·15$, significant at the 1% level.

LIMITATIONS OF THE TECHNIQUE

As long as the assumptions described on p. 53 are borne in mind, χ^2 is a very useful technique. We note here that it may be possible to combine categories in a sensible way to ensure fulfilment of the assumption relating to the size of expected frequencies. The test suffers from two difficulties when used with spatial data. First, there is an aggregation problem related to the so-called 'ecological fallacy'. The categories in our analysis are often defined arbitrarily, representing the aggregation of a number of smaller classes or areas as in the case of much census data used by human geographers, and a striking example in physical geography is given in Seddon (1971, p. 15).

Each cell frequency may therefore conceal considerable variation between its constituent subcells. Conclusions drawn at the level of analysis apply only at that level. Any assumption that they necessarily hold at a more disaggregate level, particularly at the level of the individual persons or other entities concerned, is called the **ecological fallacy**. A significant positive association between growth in the number of manufacturing establishments and city size (see Question 3, p. 63) does not mean that the number of establishments increased in *all* counties in or near large cities, only that such counties tend to be characterised by growth rather than decline. Because the associations revealed may vary with the size of areal units employed, the aggregation problem is closely associated with that of **modifiable units**. For comments on this relationship, see Haggett *et al* (1977, pp. 348—51). The second problem arises because of the non-spatial representation of spatial data; very different spatial patterns can yield identical contingency tables, and a strong spatial association may show up as a weak relationship in a contingency table. Indeed, one of the problems of many statistical techniques employed by geographers is that they involve 'throwing the spatial information away'.

Finally, it must be stressed that analysis of a problem does not end with the calculation of a test statistic. This is true of *all* statistical analysis. For instance, even if every county in or near a large city showed manufacturing growth over a given period, it is still up to the investigator to find out *why* the association occurs.

EXERCISES

1. The 31 Greater London Boroughs (the City of London was omitted because it does not hold elections on a basis comparable to that in other boroughs) were divided into three groups according to their 'political complexion'. This was done because no data were readily available of incomes or socio-economic groupings in the boroughs. Boroughs with at least twice as many Conservative as Labour councillors were classed as 'middle class', those with at least twice as many Labour as Conservative councillors as 'working class', and the remainder as 'mixed'. Definitions of this kind are not entirely satisfactory, but are often forced upon the investigator because of data limitations.

   Is there a significant difference in the suicide rates between the three types of borough? The data for 1974 are given in the table below.

Class	Population aged 15 or over	Number of suicides Observed	Number of suicides Expected
Working	2342 600	319	291
Mixed	1675 300	187	209
Middle	1966 200	239	245
	5984845	745	745

   (a) State the null and alternative hypotheses, and calculate the expected numbers of suicides using the estimate of P, $p = 0.0001245$.
   (b) Carry out the χ^2 test and interpret the results carefully.

2. Data on monthly rainfall for five stations in West Africa were obtained for July, August and September over the period 1905–73 (Bunting *et al* 1975). As a first attempt at predicting the combined August and September rainfall totals, given July rainfall totals, the observations were divided into terciles (i.e. into three groups of equal size) and entered in a 3 x 3 contingency table. Observations in the lower, middle and upper terciles were described as 'dry', 'normal' and 'wet' respectively.

July	August and September Wet	August and September Normal	August and September Dry	
Wet	8	11	4	23
Normal	11	4	8	23
Dry	4	8	11	23
	23	23	23	69

   (a) Does there appear to be a strong positive or negative association between the rainfall totals? Do this simply by inspecting the table.
   (b) Carry out a test for association, and interpret the results.

$e = 7.667$

3. Cohen and Berry (1975, p. 70) examined the relationship between initial relative accessibility to markets as measured by population potentials, and change in the total number of manufacturing establishments, for 3102 American counties. The table of observations is given below.

Manufacturing change	Accessibility of markets				
	Low	Medium	High	*e*	
Decline	297 *214.589*	304 *287.145*	268	869	*367.266*
Growth	469 *551.411*	721 *737.855*	1043	2233	*943.734*
	766	*1025*	*1311*	*3102*	

   Is there a positive or negative association between growth and accessibility? Can it simply be explained away by chance?

4. Any piece of vegetation may be characterised in terms of its life-form, and a comparison of the frequencies of five different life-forms in the Great Smoky Mountains National Park for the hardwood forests of the valleys and the higher subalpine belt produced the table below.

Altitudinal region	Life-form					Totals
	Phanero-phytes	Chamae-phytes	Hemicryto-phytes	Cryto-phytes	Thero-phytes	
Hardwood forests	41 *28·66 / 29*	5 *3*	34 *56*	29 *22*	4 *3*	113
Subalpine belt	64 *76·*	7 *9*	170 *148*	52 *59*	8 *9*	301
Totals	105	12	204	81	12	414

phanerophytes: buds relatively exposed to prevailing weather conditions.
chamaephytes: buds above ground, but at a lower level.
hemicrytophytes: buds in surface layer of soil.
crytophytes: buds are always beneath soil surface.
therophytes: plants survive adverse conditions as seeds.

(Data and definitions adapted from Seddon (1971, pp. 96—7 and 127).)

Are the profiles or histograms representing the frequency distributions of the species in the two regions significantly different?

5. Is there any evidence from the figures given below that mothers living in what are regarded as less affluent areas in Inner London are more likely to have still-born babies than those in Outer London?

Region	Birth status		Totals
	Live-born	Still-born	
Inner	41 600 *41593*	485 *492*	42 085
Outer	72 098 *72105*	859 *852*	72 957
Totals	113 698	1344	115 042

Why should you be extremely careful about interpreting the result?

REFERENCES AND FURTHER READING

In human geography use of the χ^2 test in a natural sampling context has been much favoured in studies of industrial location, as in:

Cohen, Y. S. and B. J. L. Berry 1975. *Spatial components of manufacturing change*. Chicago: University of Chicago, Dept of Geography, Research Paper no. 125.

Henderson, R. A. 1974. Industrial overspill from Glasgow: 1958–1968. *Urban Stud.* 11(1), 61–79.

White, R. L. and H. D. Watts. The spatial evolution of industry: the example of broiler production. *Trans Inst. Br. Geogs* (New Series) 2(2), 175–91.

For an example related to central place theory, see:

Fingleton, B. 1975. A factorial approach to the nearest centre hypothesis. *Trans Inst. Br. Geogs* 65, 131–40.

Various uses of the test to study plant associations and distributions of plants and vegetation types between regions are given in:

Greig-Smith, P. 1964. *Quantitative plant ecology*. London: Butterworth.

For other studies of interest to the physical geographer, see:

Bunting, A. H., M. D. Dennett, J. Elsten and J. R. Milford 1975. Seasonal rainfall forecasting in West Africa. *Nature* 253 (5493), 622–3.

So, C. L. 1974. Some coast changes around Aberystwyth and Tanybwlch. *Trans Inst. Br. Geogs* 62, 115–28.

Walling, D. E. and B. W. Webb 1975. Spatial variation of river water quality: a survey of the River Exe. *Trans Inst. Br. Geogs* 65, 155–71.

The restriction of the χ^2 test to analysis of absolute frequencies was the subject of a debate in the late 1950s between:

Mackay, J. R. 1958. Chi-square as a tool for regional studies. *Ann. Assoc. Am. Geogs* 48, 164, and

Zobler, L. 1958. The distinction between relative and absolute frequencies in using chi-square for regional analysis. *Ann. Assoc. Am. Geogs* 48, 456–7.

Further useful comments are given in:

Norcliffe, G. B. 1977. *Inferential statistics for geographers*. London: Hutchinson. (pp. 102, 110).

For methods that do not require data in this form, see:

Lewis, P. 1977. *Maps and statistics*. London: Methuen. (pp. 128–47).

Maxwell, A. C. 1961. *Analyzing qualitative data*. London: Methuen.

The problem of modifiable units is discussed in:

King, L. J. 1969. *Statistical analysis in geography*. Englewood Cliffs (New Jersey): Prentice-Hall. (pp. 154–7).

Haggett, P., A. D. Cliff and A. E. Frey 1977. *Locational analysis in human geography*, 2nd edn. Vol. 2: *Locational methods*. London: Edward Arnold. (pp. 285–90, 348–52).

The ecological fallacy is discussed in:

Robinson, W. S. 1950. Ecological correlations and the behaviour of individuals. *Am. Sociol. Rev.* 15, 351–7.

Norcliffe, G. D. 1977. *Inferential statistics for geographers*. London: Hutchinson. (Ch. 13).

For a discussion of problems of using χ^2 see:

Liebermann, B. (ed.) 1971. *Contemporary problems in statistics: a book of readings for the behavioural sciences*. New York: Oxford University Press. (pp. 279–346).

The example of the association between two plant species was taken from:

Seddon, B. 1971. *Introduction to biogeography*. London: Duckworth.

6   Relationships Over Time

INTRODUCTION

In this chapter, we first consider patterns of observations over time in terms of the distinction between systematic (dependent or deterministic) and unsystematic (stochastic) processes. A simple technique, the **runs test**, is then described which helps us to decide whether the process giving rise to a particular pattern of outcomes is likely to have been independent and random, or characterised by some kind of dependence. Finally, the **binomial theorem** and associated **binomial probability distribution**, used to calculate the probability of a given number of outcomes of a particular kind in a specified number of time periods, are introduced.

RANDOM AND SYSTEMATIC PATTERNS OVER TIME

A temporal series of outcomes may be produced by a purely independent random process. Suppose that $P(A)$ is the probability of rainfall greater than or equal to the average occurring in any given year. If rainfall conditions in any given year are not influenced by the conditions occurring in any preceding year, then

$$P(A \mid 0) = P(A)$$

where 0 represents rainfall conditions in any previous year or set of years. Turning again to our urn model, let A represent a 'wet year' (rainfall greater than or equal to the average) and B a 'dry year'. Assuming the probability of occurrence of A in any given year to be $p = 0 \cdot 5$, sampling with replacement from an urn containing ten balls, five of which are green (= wet) and five of which are red (= dry), produces a sample realisation of an independent random process operating over time. The pattern of A's and B's shown as series 1 in Figure 6.1a can be taken to represent a 50-year series of wet and dry years. There happened to be 28 wet years in this particular sample realisation, so that if we had not known the true probability to be $0 \cdot 5$, then the probability of a wet year estimated from this particular sequence would have been $28 \div 50$ or $0 \cdot 56$ which, although not correct, is not too far out. The realisation labelled series 2 in Figure 6.1a yielded 24 A's and a slight underestimate of $0 \cdot 48$.

(a)

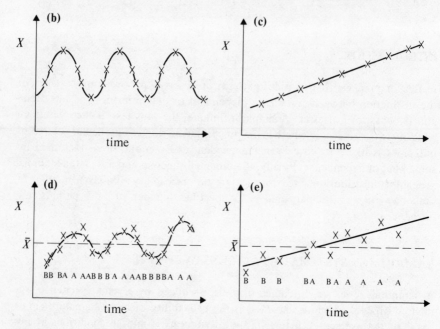

Series 1 ABAABBBABBAAAABBABABABAABABBAAAAAABABBAAAAABBABBBA

Series 2 BBABAABBABBAABAAAAAABBABBAABABBBABBBABBAABBBAABAAB

Figure 6.1   (a) Realisations of processes acting over time. (b) and (c) Examples of outcomes of deterministic processes over time. (d) and (e) Examples of outcomes of dependent and random processes over time. 'A' indicates values equal to or above $\bar{X}$, whilst 'B' indicates those below $\bar{X}$.

RANDOM NUMBER TABLES

Rather than continuing to use a physical representation of the urn model, let us make use of a **random number table**, in which each digit is generated by a chance or random procedure similar to those already described. Normally, the digits appearing in the table may consist of any of the numbers 0 to 9 inclusive. This is true of the digits shown in Table 6.1 (this has been taken from the full version given in Table A.7). Each digit might be the result of randomly drawing, with replacement, from an urn containing 10 balls labelled 0 to 9, or of spinning a roulette wheel which allows only scores from 0 to 9. Since every digit has the same probability of 'coming up' each time, they are produced by an independent random process. These numbers may be readily substituted for those we would otherwise have to obtain in a more cumbersome fashion. It is customary to start

Table 6.1   Random numbers (from Table A.7).

68327	00067	17487	49149	25894	23639	86557	04139	10756	76285
55888	82253	67464	91628	88764	43598	45481	00331	15900	97699
84910	44827	31173	44247	56573	91759	79931	26644	27048	53704
35654	53638	00563	57230	07395	10813	99194	81592	96834	21374
46381	60071	20835	43110	31842	02855	73446	24456	24268	85291
11212	06034	77313	66896	47902	63483	09924	83635	30013	61791
49703	07226	73337	49223	73312	09534	64005	79267	76590	26066
05482	30340	24606	99042	16536	14267	84084	16198	94852	44305
92947	65090	47455	90675	89921	13036	92867	04786	76776	18675

at a randomly chosen point (closing your eyes and stabbing at the page with your finger is good enough) in Table 6.1, to ensure that the sets of random numbers obtained for different purposes are all different. Otherwise, we produce patterns which are random, but identical.

> Generate two sets of 50 outcomes assuming $p = 0.2$. Let the digits 0 and 1 represent a wet year (A) and all the others a dry year (B).

The total possible number of distinctly different sample patterns of outcomes that could conceivably be generated in this way by an independent random process, assuming 50 outcomes in each pattern, is 2^{50} or approximately 1·3 million billion! There is a chance that the process *might* produce, at one extreme, 50 successive wet years and, at the other, 50 successive dry years. Notice once again the hallmark of a chance or stochastic process – the same process can produce a very large number of different realisations.

A purely dependent or deterministic sequence is generated if $p = 0.0$ or $p = 1.0$ or, if we stipulate that the first year in a series of 50 is wet (= A) and that $P(A \mid B) = 1.0, P(B \mid A) = 0.0$.

> Write down the sequence obtained if these rules are applied. What rules would be needed to generate the series AABBAABB . . . ?

In such cases, the potential number of different realisations is one. Although we cannot be absolutely sure, the realisations shown by the series of crosses in Figures 6.1b and 6.1c are very likely to have been produced by a deterministic process.

There would be no difficulty in distinguishing systematic components if real-world phenomena could be regarded as outcomes of purely dependent processes. However, most phenomena of interest to geographers are neither perfectly independent and random, nor are they perfectly dependent. For instance, rainy periods may tend to persist for four or five days before clearing, so that the probability of rain will be higher if rain fell the previous day. Letting $P(A \mid A)$ denote the conditional probability of rain today, given that it rained yesterday

$$P(A \mid A) > P(A)$$

and it would also be true that $P(A \mid B) < P(A)$, where B denotes 'no rain'. As expected, both sets of relationships between probabilities fail to conform with the definition of independence in Equation 5.1.

Generate two realisations, each of 50 outcomes, according to the rules $P(F \mid F) = 0\cdot8$ and $P(F \mid N) = 0\cdot2$, where F represents a floodyear, N a non-floodyear, and the first year in the series is a floodyear.

Other examples of dependencies in time-series data are shown in Figures 6.1d and 6.1e. In each case A denotes an 'equal to or above average outcome' and B a 'below average outcome'. The observations in Figure 6.1d might represent monthly prices of fresh fruit and vegetables in the United Kingdom, and those in Figure 6.1e the annual number of tonne kilometres of road freight traffic over a 10-year period. The systematic pattern of peaks and troughs in fresh-food prices can be ascribed to variations in the ability of home producers to supply such foods in season, and the steady (i.e. systematic) increase in road freight traffic reflects overall economic growth and the increasing diversion of goods from rail to road. As the observations are not entirely dependent or predictable, a best-fit curve or line could be drawn as shown by the broken lines (details of simple line-fitting procedures are given in Ch. 15). The line represents the systematic or predictable component and the deviations of the observations about it the unpredictable, unsystematic or random component. Fresh-food prices may happen to lie above the general trend because of bad weather in one particular year, or fall below because of a (no doubt!) temporary increase in the value of the pound in another.

THE RUNS TEST FOR RANDOMNESS

The runs test may be used to answer questions such as: Do the annual deviations about a rainfall average computed for the standard period 1936–55 for the Kano rainfall data (Ch. 2) show any evidence of climatic change over the period 1945–74 (Fig. 6.2a)? Is there any evidence of systematic trends in the number of flood days per year on the Thames at Teddington over the period 1911–40 (Fig. 6.2b)?

(a) Kano rainfall   AABBBABAAAABAAAAAABBABABBBBBBB

(b) Thames floods   BBBBAABBABBBBAAAAAAABBABBAABAA

Figure 6.2   Runs in Kano rainfall and Thames flood data.

Theory

The sets of observations in Figure 6.2, expressed in terms of binary (nominal) variables, provide typical inputs to the runs test. If, for simplicity, we consider only the last 10 years' observations of Thames floods, the pattern is

BBABBAABAA

where A denotes at or above the median flood level for the entire 30-year period 1911–40, and B below median flood level. We may compare the order in which the five A's and five B's actually occurred with all the possible orders in which they could have occurred. There are 252 possible orderings (Equation 6.2 is a formula for this value), some of which are shown in Table 6.2 ranked according to the number of **runs** they display. A run is defined by a number of consecutive outcomes of the same kind so that the runs in the flood days sequence are

BB A BB AA B AA
1   2   3   4   5   6

A run may consist of a single outcome (e.g. runs 2 and 5 above), and there are six in all. The number of runs in a series of observations can be regarded as an index of its predictability or of the amount of systematic variation it displays. The smallest possible number of runs in the flood series is two, and there are two ways in which these could have occurred (Table 6.2). Such realisations may be regarded as 'extreme', perhaps reflecting the work of processes which tend to produce a cyclic pattern about the average of the kind previously shown in Figure 6.1d. Another form of extreme is given by the two realisations at the foot of Table 6.2, each consisting of 10 runs, the maximum possible in this case. The investigator should perhaps look for a process likely to yield a highly predictable series of alternating outcomes here. 'Less extreme' realisations are found in the body of Table 6.2. Any of the 'extremes' *could* have been produced by a process similar to that of drawing, *without replacement*, five green (= A) and five red (= B) balls in random order from an urn. Notice that the sampling must be without replacement, to ensure that the sequence consists of five A's and five B's. Sampling with replacement might well produce realisations that do not have the required number of A's and B's.

Table 6.2   Listing of some of the 252 possible realisations of a randomly arranged sequence, given $n_1 = n_2 = 5$, and $n = 10$.

Realisation	Number of runs	Number of patterns showing a given number of runs
AAAAABBBBB BBBBBAAAAA	2	2
AAAABBBBBA AAABBBBBAA AABBBBBAAA ABBBBBAAAA BBBBAAAAAB BBBAAAAABB BBAAAAABBB BAAAAABBBB	3	8
. . . . .	4 5 6 7 8	. . .
ABBABABABA ABABBABABA ABABABBABA ABABABABBA BAABABABAB BABAABABAB BABABAABAB BABABABAAB	9	8
ABABABABAB BABABABABA	10	2

> Generate six random sequences of five A's and five B's. This is very easy to do using a physical representation of an urn model. How would you do it using random number tables? Where do the sequences you have generated fit in Table 6.2?

The crucial question now is: How extreme does any particular realisation have to be for us to believe it was not produced by a purely random process, but one that includes a systematic component?

As for the frequencies of outcomes in different categories in Chapter 5, the answer offered here is probabilistic. Because there are 252 different ways of ordering the five A's and five B's, the probability of obtaining two runs purely by chance is, from Table 6.2:

$$P(2 \text{ runs}) = 2/252 = 0 \cdot 008$$

and the probability of obtaining 10 runs is

$$P(10 \text{ runs}) = 2/252 = 0 \cdot 008$$

By the addition law, the probability of obtaining a pattern extreme enough to contain *either* two runs *or* tens runs must be

$P(2 \text{ runs or } 10 \text{ runs}) = P(2 \text{ runs}) + P(10 \text{ runs}) = 0 \cdot 016 \ (1 \cdot 6\%)$

The entire **probability distribution of the number of runs** can be built up from a version of Table 6.2 that shows all 252 outcomes, grouped according to the number of runs. Figure 6.3 shows the probability distribution for a series containing $n_1 = 5$ outcomes of one kind, $n_2 = 5$ outcomes of another kind, and $n = n_1 + n_2 = 10$ outcomes in all. Notice that because we can only have whole numbers of runs, the probability distribution is **discrete**, in contrast to the continuous χ^2 distribution.

To perform the runs test on our set of ten observations, first set up the hypotheses

H_0: The series is random
H_1: The series is non-random

Defining a significance level of $\alpha = 0 \cdot 05$ (5%), does the observed number of runs fall within the region of rejection? Because two possible kinds of extremes, represented by very low and very high numbers of runs, can be obtained, the region of rejection covers both tails of the probability distribution shown in Figure 6.3. Under these circumstances, the runs test is **two-tailed**.

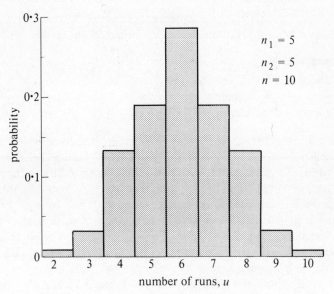

Figure 6.3   Probability distribution of number of runs.

We know already that six runs do not fall within the most extreme 1·6% of all possible randomly generated outcomes. If the probabilities of three runs and nine runs are also included, we find

$$P(3 \text{ runs or } 9 \text{ runs}) = 0·064$$

and the probability of two, three, nine or ten runs is $0·016 + 0·064 = 0·08$ or 8%. So six runs do not fall even within the most extreme 8% of all runs, and H_0 cannot be rejected at the 5% significance level. Because the probability distribution of the number of runs is discrete, the critical points (one in each tail) corresponding to $\alpha = 0·05$ cannot be determined exactly. In this case the critical numbers of runs for $\alpha = 0·05$ would be two and ten. We cannot include any more patterns to make up the probability to exactly 0·05 as this would mean including only some of the 16 realisations yielding three or nine runs, and there is no satisfactory way of deciding that any one of these realisations is more extreme than any of the others. We must include all or none, and including all would breach the 5% limit.

From the above, we should conclude that there is insufficient evidence to reject the null hypothesis of randomness and, by implication, independence.

EXAMPLE

Applying the test to the entire set of observations for the Thames given in Figure 6.2b:

$$H_0: \text{The sequence is random}$$
$$H_1: \text{The sequence is non-random}$$

and the sequence is

1	2	3	4	5	6	7	8	9	10	11	12
BBBB	AA	BB	A	BBBB	AAAAAAA	BB	A	BB	AA	B	AA

where $u = 12$ (number of runs), $n_1 = 15$ (number of years with flood-days at or above median), and $n_2 = 15$ (number of years with flood-days below median).

Let us set our significance level at $\alpha = 0·05$. Table A.2b gives upper critical values (u_U) corresponding to $\alpha = 0·05$. For $n_1 = 15$, $n_2 = 15$, this value is $u_U = 22$. Table A.2a gives lower critical values (u_L), and for $n_1 = 15$, $n_2 = 15$, this value is $u_L = 10$. If the observed number of runs, u, is greater or equal to u_U, or less than or equal to u_L (i.e. if $u \geqslant u_U$ or $u \leqslant u_L$) reject H_0 at the 5% significance level. Otherwise do not reject H_0. Since $u_L < u < u_U$, H_0 cannot be rejected at the 5% level.

If a critical value is not given in one of the Tables A.2a or A.2b, it means that any extreme number of runs cannot be regarded as significant. For instance, if $n_1 = 3$, $n_2 = 17$, then no value for u_U is given in Table A.2a. This is because the probability distribution of the number of runs is so strongly left-skewed that inclusion of the highest possible number of runs will take the probability above the designated significance level of $\alpha = 0.05$.

Construct the probability distribution of the number of runs for a five-year period with three years in which road accident rates increased (+), and two years in which they decreased (−), i.e. $n_1 = 3$, $n_2 = 2$. What is the lowest significance level (α) at which the runs test might be applied here? Do you think this satisfactory?

The test has a number of limitations. The observations must be reduced to binary form unless naturally in that state, and so information is lost. If the number of outcomes is small, no meaningful test of significance can be carried out, and Table A.2 cannot be used if the number of outcomes of any given kind exceeds 20. However, in the latter case, an approximation to the probability distribution of the number of runs is given by the **normal distribution** (see Ch. 7) and Figure 6.3 shows, even for $n_1 = n_2 = 5$, an approximation to a normal distribution (see Fig. 7.3).

PROBABILITIES OF INDEPENDENT EVENTS OVER TIME

If events are statistically independent over time, or one-dimensional space, the **multiplication law** (p. 57) may be applied. If F and N represent floodyears and non-floodyears, and p is the probability of at least one flood in any given year, then the probability of two successive floodyears is

$$P(\text{F and F}) = p \times p = p^2$$

Letting the probability of a non-floodyear be $q = (1 - p)$, the probability of two successive non-floodyears is

$$P(\text{N and N}) = q \times q = q^2$$

If $p = 0.1$, $P(\text{F and F}) = 0.1 \times 0.1 = 0.01$, or one chance in a hundred. According to the relative frequency concept of probability, of 1000 sets of two successive years, approximately 100, i.e. 0.1 or $1/10$ of 1000, should experience a flood in the first year. Assuming independence, approximately 10, i.e. 0.1 or $1/10$ of these 100, will in turn be immediately followed by another floodyear.

Assuming $p = 0.1$, calculate the probability of (a) two successive non-floodyears, (b) a floodyear followed by a non-floodyear, (c) a non-floodyear followed by a floodyear, (d) one floodyear in two successive years.

Table 6.3   Probabilities of realisations of an independent random process over time.

		$p = 0.5$	$p = 0.1$
(a) Floodyear followed by two non-floodyears (FNN)	$p \times q \times q = pq^2$	$0.5 \times (0.5)^2 = 0.125$	$0.1 \times (0.9)^2 = 0.081$
(b) Two non-floodyears followed by floodyear (NNF)	$q \times q \times p = q^2 p$	$(0.5)^2 \times 0.5 = 0.125$	$(0.9)^2 \times 0.1 = 0.081$
(c) Five years without floods (NNNNN)	$q \times q \times q \times q \times q = q^5$	$(0.5)^5 = 0.03125$	$(0.9)^5 = 0.59049$
(d) Ten years without floods (NNNNNNNNNN)	q^{10}	$(0.5)^{10} = 0.00098$	$(0.9)^{10} = 0.34868$
(e) Floods every year for five years (FFFFF)	p^5	$(0.5)^5 = 0.03125$	$(0.1)^5 = 0.00001$

The multiplication law can be readily extended to deal with more than two consecutive outcomes, simply by multiplying together three individual probabilities for three outcomes, ten for ten outcomes and, in general, k probabilities for k outcomes. Some examples are given in Table 6.3. Civil engineers, farmers and insurance companies are interested in the kinds of probabilities labelled (c) and (d). A cultivated zone in a river valley may be worth extending if there are no floods during the next five years. If $p = 0.5$, then the chance of such an occurrence is only 0.03125 or about 3 in 100. No farmer is likely to take such a risk, and even if he did, insurance companies would show little interest in supporting him. Such a probability might persuade the flood engineer, and the appropriate authorities, to initiate a flood control programme in the area. If $p = 0.1$, some farmers might well be prepared to take the risk, since there is a better than even chance (0.59 or 59%) of no flood occurring during the critical period. Extension of the critical period to ten years ((d) in Table 6.3) brings different probabilities to be appraised. Reactions of farmers, engineers and insurance companies may also depend on other factors. Nevertheless, probabilistic thinking may form an important element of those reactions, as described by Kates (1962) who compares farmers and engineers, and Newson (1975) and Weyman (1975) who illustrate the views of engineers and hydrologists.

> If $p = 0.3$, show that the probability of one floodyear in two successive years is 0.42. In order to do this calculation, what factor must you take into account that is not illustrated in Table 6.3? Show also that the probability of two floodyears in three successive years is 0.432 if $p = 0.6$.

More complex probabilities of the kind considered in the foregoing exercise are readily calculated if we make use of the **binomial theorem**.

THE BINOMIAL THEOREM AND THE BINOMIAL PROBABILITY DISTRIBUTION

Assuming independence over time, and that one of only *two* possible outcomes can occur (hence the term *bi*-nomial) in any given time interval of a series, the binomial theorem allows us to calculate the probability that any specified number of time intervals is characterised by one of the outcomes. Suppose we are interested in the number of drought years likely to occur in the next five. Many of the patterns that might occur are given in Table 6.4, D denoting drought, and N non-drought. To ascertain, for example, the probability of three drought years in five, we must establish how many different ways three drought years could possibly occur in five. Table 6.4 shows there are ten possible ways in all, each of these ways being a district arrangement known as a **combination**. If $p = P(D)$, $q = 1 - p = P(N)$, the probability of any particular combination of k drought years in n can be found by the multiplication law as

$$p^k q^{n-k} \qquad\qquad (6.1)$$

Applied to our problem, this gives $p^3 q^2$, and as there are ten combinations of

Table 6.4   Derivation of binomial probabilities.

Combinations or arrangements	Number of drought years (k)	Number of combinations	Probability of k drought years
NNNNN	0	1	$\binom{5}{0}p^0 q^5$
DNNNN NDNNN NNDNN NNNDN NNNND	1	5	$\binom{5}{1}p^1 q^4$
?	2	?	?
NNDDD DNNDD DDNND DDDNN NDNDD NDDND NDDDN DNDND DNDDN DDNDN	3	10	$\binom{5}{3}p^3 q^2$
NDDDD DNDDD DDNDD DDDND DDDDN	4	5	$\binom{5}{4}p^4 q^1$
DDDDD	5	1	$\binom{5}{5}p^5 q^0$

three in five

$$P(X = 3) = 10p^3q^2$$

where X represents the number of drought years.

Rather than enumerate combinations by writing them all out each time, the number may be obtained from the formula for a **binomial coefficient** $\binom{n}{k}$ where

$$\binom{n}{k} = \frac{n!}{(n - k)!k!} \tag{6.2}$$

The symbol ! denotes 'factorial' and $n!$ (n factorial) $= n \times (n - 1) \times (n - 2) \times \ldots \times (3) \times (2) \times (1)$. It is always assumed that k is a positive whole number and $k \leqslant n$. The number of combinations of $k = 3$ droughts in $n = 5$ successive years is therefore

$$\binom{5}{3} = \frac{5!}{(5 - 3)!3!} = \frac{5!}{2!3!} = \frac{5 \times 4 \times 3 \times 2 \times 1}{(2 \times 1) \times (3 \times 2 \times 1)} = 10$$

and confirms the answer previously obtained. We can thus use the notation $\binom{n}{k}$ to indicate the number of ways of arranging k items in a total of n. Since we also know the probability of each is $p^k q^{n-k}$, it follows that the total probability is

$$P(X = k) = \binom{n}{k} p^k q^{n-k} \tag{6.3}$$

This gives the general formula for a **binomial probability**. When computing values of $\binom{n}{k}$ observe the conventions that $0! = 1$, and that any number raised to the power zero equals 1. For example, if $p = 0{\cdot}3$, the probability of no drought years in five is

$$P(X = 0) = \binom{5}{0} \times (0{\cdot}3)^0 \times (0{\cdot}7)^5 = \frac{5!}{5!0!} \times (1) \times (0{\cdot}7)^5$$

$$= 1 \times 1 \times 0{\cdot}1681 = 0{\cdot}1681$$

It can also save work to remember that

$$\binom{n}{k} = \binom{n}{n - k} \qquad \text{e.g.} \quad \binom{5}{3} = \frac{5!}{2!3!} = \binom{5}{2}$$

because the number of ways in which three drought years can occur in five must be the same as the number of ways two non-drought years can occur in five.

> Write out the missing combinations of two drought years in five in Table 6.4. Make sure the number you find agrees with your calculation of $\binom{5}{2}$ in the expression for the binomial probability for this case, also missing from the table.

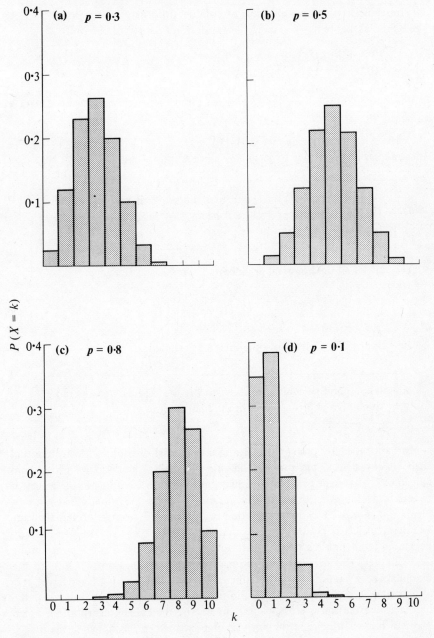

Figure 6.4   Examples of binomial distribution with $n = 10$ and varying values of p.

Probabilities of all possible sets of combinations may be computed if, using the binomial theorem, we expand the expression $(p + q)^n$ to obtain

$$\binom{n}{0}p^0 q^n + \binom{n}{1}p^1 q^{n-1} + \binom{n}{2}p^2 q^{n-2} + \ldots + \binom{n}{n-2}p^{n-2}q^2 +$$

$$\binom{n}{n-1}p^{n-1}q^1 + \binom{n}{n}p^n q^0 \tag{6.4}$$

the successive terms giving $P(X = 0), P(X = 1), \ldots, P(X = n - 1), P(X = n)$. The first and last terms reduce to q^n and p^n since $p^0 = q^0 = 1$ and $\binom{n}{0} = \binom{n}{n} = 1$.

> Show that $\binom{n}{0} = \binom{n}{n} = 1$, and that the series of probabilities in the final column of Table 6.4 behaves in accordance with the binomial theorem.

The sum of all the binomial probabilities should be $1 \cdot 0$, i.e.

$$\sum_{k=0}^{n} \binom{n}{k}p^k q^{n-k} = 1 \cdot 0$$

because all possible patterns of outcomes are included.

> Use a hand calculator to show this is the case by working out the binomial probabilities in the final column of Table 6.4.

The entire set of probabilities forms a **binomial distribution**, whose overall shape is revealed if the probabilities are plotted in the form of a histogram (Fig. 6.4a). The shape of a binomial distribution varies according to the values of p and n. Holding n constant, and varying p, gives the other histograms of Figure 6.4. For $p < 0 \cdot 5$ the distribution is right-skewed, for $p > 0 \cdot 5$ left-skewed, and symmetrical if $p = 0 \cdot 5$. Notice that for $p = 0 \cdot 5$ the distribution is already beginning to resemble a normal distribution (see Fig. 7.3). As we shall see in the next chapter, the normal distribution provides a good approximation to the binomial for $p = 0 \cdot 5$ and n greater than 15 or 20, or when n is 'very large'. It is also true that if n is 'very large', and p relatively small — say $0 \cdot 1$ or less — the binomial distribution is closely approximated by the **Poisson distribution**, which plays an important role in the analysis of point patterns as shown in Chapter 8.

Mean and standard error of the binomial distribution

If X is a variable, e.g. number of drought years, which follows a binomial distribution, then X has mean or expected value

$$\mu = E(X) = np \tag{6.5}$$

and standard error

$$\sigma = \sqrt{npq} \qquad\qquad (6.6)$$

For $n = 5$, $p = 0.3$, the expected number of drought years is $5 \times 0.3 = 1.5$, and the standard error is $\sqrt{5 \times 0.3 \times 0.7} = 1.0247$. If we consider a very large number of five-year periods in an area where $p = 0.3$, the average number of years with drought should be approximately 1.5.

EXERCISES

1. A survey of premises along one side of a local high street recorded the following sequence of convenience goods (A) and comparison goods (B) shops:

   AABBBAABABBBBABAAAABAAAA.

   Do you think this sequence represents a random arrangement of the two shop types? Is your conclusion altered if *a priori* it had been hypothesised that convenience and comparison goods shops tend to be segregated and therefore show clustering?

2. Suppose that a particular plant species requires three consecutive years in which monthly winter temperatures do not fall below $-3\,^\circ$C if it is permanently to colonise an area. If the probability of a temperature below $-3\,^\circ$C in any given year is $p = 0.2$, does that species have a better than even chance of establishment?

3. Suppose that the fruiting and regeneration of certain trees requires a mean temperature of $21\,^\circ$C in the warmest month of the year, and that the lifespan of the trees is about $n = 100$ yr. This means such conditions are essential at least once a century, to ensure replacement as trees die out (and assuming the trees to be of approximately the same age − why?). If the probability of the critical temperature being equalled or surpassed in any year is $p = 0.01$, do you think regeneration likely? Is your answer at all modified if you calculate the probability of *no* year in 100 reaching the critical figure (use log tables)? How much more optimistic or pessimistic are you if $p = 0.02$, or $p = 0.005$? If the trees failed to regenerate, could you conclude that a climatic change had occurred?

REFERENCES AND FURTHER READING

Properties of the binomial distribution are well explained in:
Bulmer, M. G. 1967. *Principles of statistics*, 2nd edn. Edinburgh: Oliver and Boyd. (pp. 81−5).
A significance test based on this distribution is given in:
Norcliffe, G. B. 1977. *Inferential statistics for geographers*. London: Hutchinson. (pp. 73−81).
The role of probabilities (both 'subjective' and 'objective') on human behaviour, is discussed by:
Kates, R. W. 1962. *Hazard and choice perception in flood plain management*. Chicago: University of Chicago, Department of Geography, Research Paper no. 78.

Newson, M. D. 1975. *Flooding and flood hazard in the United Kingdom.* Oxford University Press.

Weyman, D. 1975. *Runoff processes and streamflow modelling.* Oxford: Oxford University Press.

Exercise 3 is based on an idea in:

Gregory, S. 1978. *Statistical methods and the geographer*, 4th edn. London: Longmans. (pp. 75–7).

Table 6.1 is taken from:

Neave, H. R. 1978. *Statistics tables*. London: George, Allen & Unwin.

7 The Normal Distribution

INTRODUCTION

The normal probability distribution is important for a number of reasons. First, it provides a fairly close approximation to many — but by no means all — frequency distributions of real-world observations. Secondly, it is necessary to assume that observed frequency distributions are approximately normal if some of the statistical tests widely employed in an artifical sampling context (e.g. t and F tests), and discussed in Section III of this book, are to be carried out. Thirdly, many sampling distributions are found to be approximately normal under certain conditions, specified by the **Central Limit Theorem** (Ch. 11), even if the frequency distributions of the observations from which they are derived are not themselves normal. Examples will be given with respect to the probability distribution of the number of runs and the binomial distribution, introduced in the previous chapter. Some of the statistical tests used to examine spatial point (Ch. 8) or area (Ch. 9) patterns are also based on a normally distributed test statistic. In general, the normal distribution is liable to be encountered whenever a relatively large number of (statistically) independent influences or observations (25 or more) is added together.

VARIABLES SHOWING AN APPROXIMATELY NORMAL DISTRIBUTION

Theoretically, a normally distributed variable is *continuous* and can take on any value between plus and minus infinity. In practice, many take on only a positive value (including zero), e.g. rainfall in mm, pressure in mbar, distance in km, or a positive or negative value in a fairly restricted range, e.g. temperature, height or slope angle differences. It may also be convenient to treat a variable as if it were continuous even if, strictly speaking, it is not. This can be done if the range of possible values of a variable is fairly 'large' relative to the basic unit of measurement. For example, populations of cities may take on any discrete value such as 121 157 or 6031 044 within a vast range, say from 100 000 to 10 000 000. Even a discrete variable with a range of only 20 or 30 units may be usefully regarded as continuous.

The histograms of slope-angle frequency distributions for a site near Bernalillo, New Mexico, and in the Verdugo Hills, California, shown in Fig. 7.1a, are taken from a classic study by Strahler (1950). Slope angle may be regarded as a truly continuous variable. According to Strahler, the symmetry of each distribution shows that, within a small region where conditions are more or less

(a) Slope angles

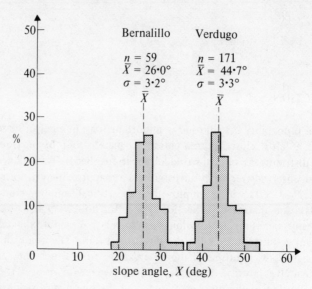

(b) Familiarity scores

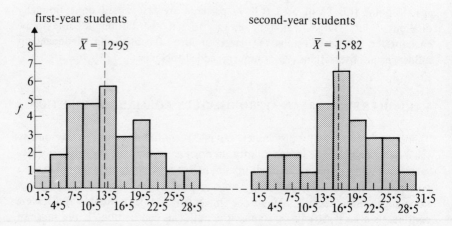

Figure 7.1   Examples of normally distributed variables.

uniform, slopes tend to approach an equilibrium angle approximated by the mean (26·0° for Bernalillo, 44·7° for Verdugo), and determined by dominant or systematic factors such as lithology, climate, soil, vegetation and relief. Fluctuations about the mean are attributed to systematic but minor variations in aspect and microclimate, and some of the supposedly dominant influences may tend to cancel out. Notice that points made in our earlier discussion of systematic and

Table 7.1   Annual natural flows on the Thames at Teddington.

(a) Frequency distribution.

Class interval	150–224	225–299	300–374	375–449	450–524	525–599	600–674	675–749	750–824	825–899	900–974
Frequency (f)	1	2	6	3	7	11	7	6	2	4	1

(b) Cumulative frequencies and percentages.

Cumulative frequency	1	3	9	12	19	30	37	43	45	49	50
Cumulative percentage	2	6	18	24	38	60	74	86	90	98	100

random sources of variation have been applied here to the special case in which the frequency distribution of measurements is normal.

> Table 7.1a gives the frequency distribution for annual natural flows of the Thames at Teddington for the period 1915–64 (inclusive). Can you think of any systematic (or dominant) and random elements which might produce such a distribution?

Although data obtained by human geographers are less likely to be normally distributed, perception and attitude studies carried out by behavioural geographers provide some of the exceptions. In a study carried out by third-year undergraduates (Ratcliff *et al.* 1977), a sample of students at Reading University were asked how familiar they felt themselves to be with scenes shown on 36 colour slides all taken within ¼ mile of Reading town centre. The minimum and maximum possible scores were 0 and 36, and feasible scores were 0, ½, 1, 1½, . . . 35½, 36, so that measurement is not, strictly speaking, on a continuous scale. Results for first- and second-year students are shown in Figure 7.1b. On average, students in both years were familiar with less than half the scenes, and it could be argued that the dominant factor at work here is the strong campus orientation of the students – most live on campus and do not visit the town centre very often. However, variation about the mean can be attributed to a variety of systematic factors such as differences in usual mode of travel to the town centre, place of residence, and in the attitude and behaviour of males and females. A more cynical view would be that most people, including students, wish to avoid appearing too brash or too ignorant, and so plump with due modesty for scores in the 10 to 20 range!

The normal distribution is sometimes known as the **normal law of errors**, as repeated measurements of the same fixed quantity tend to show a normal distribution about that quantity (Bulmer 1967, p. 108).

PROPERTIES OF THE NORMAL DISTRIBUTION

The normal curve

The **normal curve** which traces out the overall shape of the normal distribution may be described by the mathematical function

$$f(X) = \frac{1}{\sigma\sqrt{2\pi}} \, e^{-(X-\mu)^2/2\sigma^2}$$

where X is a normally distributed variable with mean μ and standard deviation σ, and π and e are the mathematical constants $3\cdot1416\ldots$ and $2\cdot7182\ldots$ respectively. Depending upon the values of μ and σ, the normal distribution may assume any variety of positions and shapes. Thus, in Figure 7.2a, the normal distributions each have the same standard deviation or shape but different means or positions, whereas in Figure 7.2b they have the same mean but different standard deviations and, hence, differing shapes. Every normal distribution is symmetrical (i.e. *not* skewed) about its mean, and its mean, median and mode are equal. Notice also how the normal curve eventually flattens out to approach the horizontal axis asymptotically (that is, it keeps approaching the axis without ever quite reaching it) as we move further away from the mean.

Areas under the normal curve

The total area under the normal curve represents a probability of $1\cdot0$ or 100%. Given the mean and standard deviation of *any* normal distribution, it can also be shown that:

(a) approximately $68\cdot26\%$ of the area under the curve lies within ±1 standard deviation of the mean, and so the probability of a value occurring within such limits is $0\cdot6826$,

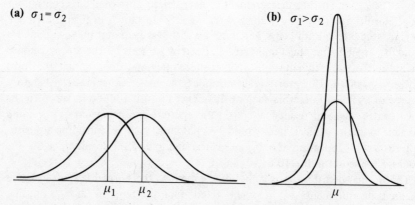

(a) $\sigma_1 = \sigma_2$ **(b)** $\sigma_1 > \sigma_2$

μ_1 μ_2 μ

Figure 7.2   Examples of normal curves.

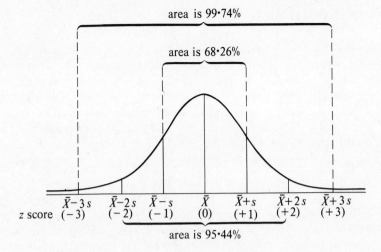

area is 99·74%

area is 68·26%

| | $\bar{X}-3s$ | $\bar{X}-2s$ | $\bar{X}-s$ | $\bar{X}$ | $\bar{X}+s$ | $\bar{X}+2s$ | $\bar{X}+3s$ |
| z score | (−3) | (−2) | (−1) | (0) | (+1) | (+2) | (+3) |

area is 95·44%

Figure 7.3   Areas under the standard normal curve.

(b) approximately 95·44% of the area lies within ±2 standard deviations of the mean, giving a probability of 0·9544, and

(c) approximately 99·74% of the area lies within ±3 standard deviations of the mean, giving a probability of 0·9974.

This is illustrated in Figure 7.3. We may apply these results to the Thames discharge data of Table 7.1a, which have a sample mean $\bar{X} = 570$ and a sample standard deviation $s = 172$, the units being thousand million gallons. (The symbols $\bar{X}$ and s are used, rather than μ and σ, because the measurements may be regarded as a *sample* realisation of a process operating over time.) Assuming these observations approximately follow a normal distribution, the probability of the discharge X, in any given year, falling between the limits $(\bar{X} - 1s) = 570 - (1 \times 172)$ and $(\bar{X} + 1s) = 570 + (1 \times 172)$ must be 0·6826, i.e.

$$P(398 \leqslant X \leqslant 742) = 0·6826$$

Show that the limits within which X falls about the mean with probability **0·9544** are **226** and **914**, and write down the limits within which X falls with probability 0·9974.

Given their mean and standard deviation, we may write down similar limits for other normally distributed variables, such as the slope and familiarity observations of Figure 7.1.

THE STANDARD NORMAL DISTRIBUTION

A method is available for obtaining the probabilities of the previous section, or any other normal probabilities in which we might conceivably be interested, and is particularly convenient because it applies to *any* normal distribution, no matter what its mean and standard deviation. The normal distribution in question is transformed into a **standard normal distribution** which has a mean of zero and a standard deviation of 1·0 (Fig. 7.3). Suppose the probability of an annual discharge between $\bar{X} = 570$ and $X = 700$ is required. The difference between X and $\bar{X}$ is expressed in standard deviation units using the formula

$$z = \frac{X - \bar{X}}{s} \qquad (7.1)$$

and is known as a z score or **standard normal deviate**. Applied to our example, this gives

$$z = \frac{700 - 570}{172} = +0·76$$

so that a value of X located 130 units above a mean of 570 before transformation is now located 0·76 standard deviation units above a mean of zero after transformation. The sign of z should be retained since positive scores always lie above the mean, and negatives ones below. The probability we require is represented by the shaded area in Figure 7.4a, and may be obtained directly from Table A.3. In the table, the figures along the top and down the left-hand side are z scores, To find the value for 0·76, read off the entry at the intersection of the 0·70 row and the 0·06 column. This is 0·2764. The probability of any value falling in the interval between 0 and 0·76 under the standard normal curve is therefore 0·2764, i.e.

$$P(0·76) = 0·2764$$

or, in terms of the original untransformed normal distribution

$$P(570 \leqslant X \leqslant 700) = 0·2764$$

Let us consider other probabilities that might be of interest. Because the normal distribution is symmetrical about its mean, it immediately follows that the probability of an annual discharge between 440 and 570 thousand million gallons is 0·2674. The value 440 is 130 units below 570, and 0·76 standard deviation units below $\bar{X}$ (Fig. 7.4b), hence

$$P(440 \leqslant X \leqslant 570) = P(570 \leqslant X \leqslant 700) = 0·2764$$

The z score would be negative, because $(X - \bar{X}) = -130$.

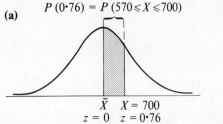

(a) $P(0·76) = P(570 \leqslant X \leqslant 700)$

$\bar{X}$ $X = 700$
$z = 0$ $z = 0·76$

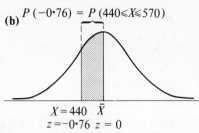

(b) $P(-0·76) = P(440 \leqslant X \leqslant 570)$

$X = 440$ $\bar{X}$
$z = -0·76$ $z = 0$

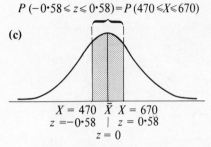

$P(-0·58 \leqslant z \leqslant 0·58) = P(470 \leqslant X \leqslant 670)$

(c)

$X = 470$ $\bar{X}$ $X = 670$
$z = -0·58$ $|$ $z = 0·58$
$z = 0$

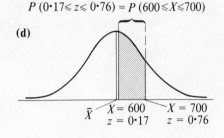

$P(0·17 \leqslant z \leqslant 0·76) = P(600 \leqslant X \leqslant 700)$

(d)

$\bar{X}$ $X = 600$ $X = 700$
$z = 0·17$ $z = 0·76$

Figure 7.4   Obtaining probabilities of normally distributed outcomes.

According to Table A.3, how much of the area under the normal curve lies between $z = 0·0$ and $z = 1·00$? Use the answer to confirm finding (a) on p. 84. Then, use Table A.3 again to confirm findings (b) and (c).

The probability of an annual discharge exceeding 700 thousand million gallons is readily found by subtraction, because half of the area under the curve lies above the mean (Fig. 7.4a), i.e.

$$P(X \geqslant 700) = 0·5000 - P(570 \leqslant X \leqslant 700)$$
$$= 0·5000 - 0·2764 = 0·2236$$

What is the probability of an annual discharge less than or equal to 440 thousand million gallons? Write out the probability symbolically in 'natural units' and in standard deviation units.

If the desired probability is for a range of values symmetrically straddling the mean, e.g. $P(470 \leqslant X \leqslant 670)$, calculate

$$z_1 = \frac{X_1 - \bar{X}}{s} = \frac{670 - 570}{172} = 0 \cdot 58$$

We find $P(z_1) = P(0 \cdot 58) = 0 \cdot 2190$. By symmetry, we know $z_2 = -0 \cdot 58$ and $P(z_2) = 0 \cdot 2190$ and so

$$P(470 \leqslant X \leqslant 670) = P(z_1) + P(z_2) = 0 \cdot 4380$$

This calculation is illustrated in Figure 7.4c.

Show that $P(470 \leqslant X \leqslant 670) = 0 \cdot 4380$, and that $P(100 \leqslant X \leqslant 600) = 0 \cdot 5643$. Draw diagrams to illustrate your answers.

Finally, refer to Figure 7.4d and show that $P(600 \leqslant X \leqslant 700) = 0 \cdot 2089$.

STATISTICAL INFERENCE

As stated in the introduction, the normal distribution is important in inferential statistics because many test statistics are normally distributed if the number of observations is sufficiently large, and because some tests of significance assume normally distributed observations (Chs. 11 and 12).

Critical points for one- and two-tailed tests at the conventional significance levels of $\alpha = 0 \cdot 05$ and $\alpha = 0.01$ may be found by following the table look-up procedure described earlier, *but in reverse*. Suppose a *two-tailed test* is being carried out at the $0 \cdot 05$ or 5% level. The critical values of z defining the region of rejection must be such that $2 \cdot 5\%$ of the area under the curve lies beyond each value. Because of symmetry, it is necessary only to find one such value of z. Table A.3 gives the area between the mean and a specified value of z, so we want the probability value in the table which, when subtracted from $0 \cdot 5000$, leaves a remainder of $0 \cdot 0250$. That value is $0 \cdot 4750$, and lies at the intersection of the $1 \cdot 90$ row and $0 \cdot 06$ column. The critical values of z, used as a test statistic at the 5% significance level, are therefore $+1 \cdot 96$ and $-1 \cdot 96$. For a significance level of $0 \cdot 01$ or 1%, the values are found to be $+2 \cdot 58$ and $-2 \cdot 58$. These values have been obtained by rounding-up, as $0 \cdot 4950$ falls midway between $0 \cdot 4949$ and $0 \cdot 4951$, corresponding to $z = \pm 2 \cdot 57$ and $z = \pm 2 \cdot 58$ respectively,

---

Find the critical values of z corresponding to the $0 \cdot 05$ and $0 \cdot 01$ levels for a one-tailed test.

Tests for normality should be carried out before any of the parametric

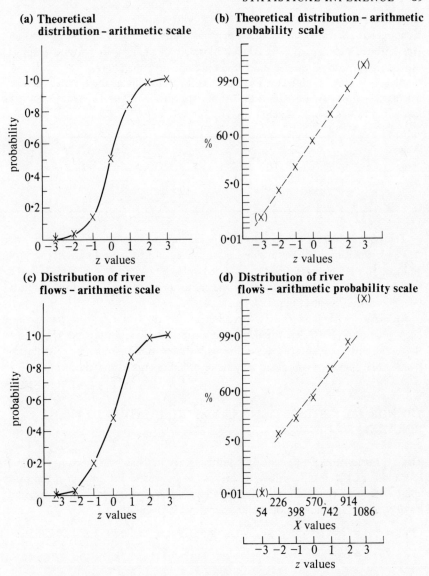

Figure 7.5   Cumulative probability for the normal distribution.

methods, based on the t and F distributions, described in Chapters 11 and 12, are employed. Several methods are available. The simplest is to plot the data in the form of a histogram, and ask whether it seems reasonable to believe that the observations have been obtained from a set of all possible observations which is normally distributed. There may also be theoretical grounds for thinking that the observations should be normally distributed, e.g. slope measurements in a region of uniform bedrock. The most rigorous method is to compare observed and expected frequencies using a goodness-of-fit test such as χ^2 or Kolmogorov—Smirnov, and details may be found in Taylor (1977, p. 128) and Weatherburn (1962), p. 175). Under most circumstances, however, a graphical procedure is quite adequate, and is the only one described here.

As a first step, cumulative probabilities for values under the normal curve are calculated. For any normal distribution, the probabilities of values occurring *below* the values specified below are given in the table.

Value	$\bar{X} - 3s$	$\bar{X} - 2s$	$\bar{X} - s$	$\bar{X}$	$\bar{X} + s$	$\bar{X} + 2s$	$\bar{X} + 3s$
z value	−3	−2	−1	0	+1	+2	+3
Cumulative probability	0·001	0·023	0·159	0·500	0·841	0·977	0·999

The probabilities are calculated by looking up the appropriate values in Table A.3 and rounding to three decimal places. When plotted, these cumulative probabilities form the characteristic S-shaped or sigmoid curve of Figure 7.5a. The vertical axis may be scaled, as on special **probability paper**, so that the curve becomes a straight line (Fig. 7.5b).

Cumulative frequencies and percentages for the Thames flow figures have been given in Table 7.1b. When plotted on special probability paper (Fig. 7.5d), the cumulative percentages show a roughly linear trend and so the distribution from which they were obtained may be regarded as approximately normal.

THE NORMAL DISTRIBUTION AS AN APPROXIMATION TO THE BINOMIAL

The expression for a **binomial probability**, given in Chapter 6, becomes difficult to handle as the total number of events, n, increases. For instance, the probability of 20 of the next 30 years experiencing above-average rainfall, assuming $p = 0·5$, and independence of events over time, is given by

$$P(X = 20) = \binom{30}{20}(0·5)^{20}(0·5)^{10} = 30\ 045\ 015 \times 0·000000954 \times 0·00976563$$
$$= 0·0280$$

a rather tedious calculation! However, we know from Formulae 6.5 and 6.6 that the mean of the binomial distribution from which this probability was obtained is $np = 30 \times 0·5 = 15·0$, and its standard deviation is $\sqrt{npq} = \sqrt{30 \times 0·25} = 2·74$. A normal distribution with the same mean and standard deviation provides a close approximation to this binomial distribution. To calculate $P(X = 20)$, some kind of subterfuge is necessary, because the probability of any *individual* value cannot be determined from a continuous distribution. However, we compute $P(X = 20)$ *as if* this represents all possible outcomes under a normal distribution between 19·5 and 20·5. The probability is obtained, first by resolving the area under the curve into two components, and then by subtraction, as illustrated in Figure

7.4d:

$$P(19\cdot5 \leqslant X \leqslant 20\cdot5) = P(15\cdot0 \leqslant X \leqslant 20\cdot5) - P(15\cdot0 \leqslant X \leqslant 19\cdot5)$$

$$z_1 = \frac{20\cdot5 - 15\cdot0}{2\cdot74} = 2\cdot01 \qquad z_2 = \frac{19\cdot5 - 15\cdot0}{2\cdot74} = 1\cdot64$$

From Table A.3, $P(2\cdot01) = 0\cdot4778$ and $P(1\cdot64) = 0\cdot4495$, and so the desired probability is

$$P(2\cdot01) - P(1\cdot64) = 0\cdot4778 - 0\cdot4495 = 0\cdot0283$$

differing by only $0\cdot0003$ from the equivalent binomial probability.

If calculating tail probabilities, i.e. of the form $P(X \geqslant a)$ or $P(X \leqslant b)$, the **correction for continuity** is employed by subtracting $0\cdot5$ from values above the mean, and adding $0\cdot5$ to values below the mean, e.g. $P(X \geqslant 18\cdot5)$ rather than $P(X \geqslant 19\cdot0)$ and $P(X \leqslant 5\cdot5)$ rather than $P(X \leqslant 5\cdot0)$ for the above example.

The approximation holds good for small values of n between 10 and 25 if $p = 0\cdot5$. When p is 'not too small' or 'not too large', lying between $0\cdot2$ and $0\cdot8$, the approximation is also reasonable for $n \geqslant 25$. For p close to $0\cdot0$ or $1\cdot0$, the approximation is poor, and the Poisson distribution used instead (see Ch. 8). As a rule of thumb, use the approximation only when np and nq are *both* greater than five.

THE NORMAL DISTRIBUTION AS AN APPROXIMATION TO THE PROBABILITY DISTRIBUTION OF THE NUMBER OF RUNS

The maximum number of items of any particular kind that can be accommodated using specially constructed tables was found to be 20 (Ch. 6), i.e. both n_1 and n_2 must be limited to 20. If either exceeds 20, it can be shown that the probability distribution of the number of runs is approximately normal with mean

$$\mu = \frac{2n_1 n_2}{n_1 + n_2} + 1 \tag{7.2}$$

and standard error

$$\sigma = \sqrt{\frac{2n_1 n_2 (2n_1 n_2 - n_1 - n_2)}{(n_1 + n_2)^2 (n_1 + n_2 - 1)}} \tag{7.3}$$

If u represents the observed number of runs, then

$$z = \frac{u - \mu}{\sigma} \tag{7.4}$$

is a standard normal deviate.

A longer series of observations on West African rainfall in the Sahel region is available for the period 1905–74, positive (or zero) deviations from the zonal average being denoted by A, and negative deviations by B:

AABAABABBBBABABABAABBAABABAAABABBABAABBBABAAABBBAB
AAAABAAAAAABBABABBBBBBBB

The number of A's is $n_1 = 35$, and of B's is $n_2 = 35$. The number of runs is given by $u = 38$. Because there is no prior information or expectation as to the form any systematic variation in this series should take, a *two-tailed test* is appropriate:

$$H_0: \text{the series is random,}$$
$$H_1: \text{the series is non-random,}$$

and the calculations are

$$\mu = \frac{2(35)(35)}{70} + 1 = 36$$

$$\sigma = \sqrt{\frac{2(35)(35)[(2 \times 35 \times 35) - 35 - 35]}{(35 + 35)^2(35 + 35 - 1)}} = \sqrt{17 \cdot 246} = 4 \cdot 153$$

and

$$z = \frac{|38 - 36|}{4 \cdot 153} = 0 \cdot 48$$

The absolute value symbol $|\ldots|$ is used because the sign of the test statistic is ignored when carrying out a two-tailed test.

> Is there strong evidence against H_0? Do you agree with Bunting *et al* (1976, p. 63) that 'the recent succession of drought years falls within the statistical expectation, and does not, in our view, indicate a changing climate in the zone that we have considered'.

EXERCISES

1. In each of the following situations, state whether you think the observations would be normally distributed or not. What are the systematic and random factors at work?
   (a) Catastrophic floods.
   (b) Distances between households and bread shops within a city.
   (c) Distances between households and a single hypermarket on the edge of a city.

(d) Driving times between households and a single hypermarket on the edge of a city (assuming the city has a motorway-standard by-pass).
(e) Percentage loss of weight on ignition for a large number of soil samples obtained from a similar soil type.
(f) Orientations of stones sampled from a glacial deposit.

2. A country has already experienced three years in succession with annual rainfall below 500 mm. Analysis of records shows no sign of statistical dependence in rainfall figures over time. If mean annual rainfall is $\overline{X} = 800$ mm, with standard deviation $s = 150$ mm, what is the probability of three further years with rainfall below 500 mm? If six successive years with rainfall below 500 mm occurred, how might you justify a view that a climatic change has occurred?

3. The probability of rain on any given Saturday afternoon during the football season in Manchester is $p = 0\cdot5$. Assuming 30 home games per season, what is the probability of being soaked
(a) less than five times this season,
(b) more than 20 times,
(c) less than 15 times,
(d) between 10 and 20 times, and
(e) only once?

4. The sequence of annual natural flows of the Thames at Teddington 1915–64 is shown below, where A represents equal to or above mean annual flow, and B below mean annual flow:

AAABABBBAAAAAABAAABBBAABAAABBBBAABBBABBABBBABA
ABBB

(a) Is use of the mean as a cut-off point justified here?
(b) Is there any significant trend in the observations over time?

REFERENCES AND FURTHER READING

The normal law of errors is discussed in:
  Bulmer, M. G. 1967. *Principles of statistics,* 2nd edn. Edinburgh: Oliver and Boyd. (Ch. 7).
Good examples of normally distributed variables may be found in:
  Carson, M. A. 1971. An application of the concept of threshold slopes to the Laramie Mountains, Wyoming. In *Slopes: form and process,* D. Brunsden (ed.), 31–48. Institute of British Geographers, Special Publication no. 3.
  Strahler, A. H. 1950. Equilibrium theory of erosional slopes approached by frequency distribution analysis. *Am. J. Sci.* **248**, 673–96, 800–14.
See also:
  Chorley, R. J. 1966. The application of statistical methods in geomorphology. In *Essays in geomorphology,* G. H. Dury (ed.). London: Heinemann. (And the references therein.)
For the use of z scores to ensure comparability between regions or groups, see:
  Smith, D. M. 1977. *Patterns in human geography.* London: Penguin. (pp. 154–7).
The data on student knowledge were taken from:
  Ratcliff, H. M., A. P. Pilditch and N. R. Walter 1977. *A study of students' knowledge of Reading town centre.* University of Reading, Dept of Geography, Unpublished Behavioural Geography Project.

Rainfall trends are examined in:

Bunting, A. H., M. D. Dennett, J. Elsten and J. R. Milford 1976. Rainfall trends in the West African Sahel. *Q. J. R. Meteorol. Soc.* **102**, 59–64.

For goodness-of-fit tests for normality, see:

Taylor, P. J. 1977. *Quantitative methods in geography.* Boston: Houghton Mifflin.

Weatherburn, C. E. 1962. *A first course in mathematical statistics,* 2nd edn. Cambridge: Cambridge University Press.

8   Relationships Over Space I: Point Patterns

INTRODUCTION

Geographers have long been interested in the distribution of phenomena in space as represented by points on dot maps, whether they be, on the one hand, settlements, dwellings or retail outlets, or, on the other hand, limestone solution hollows, drumlins or inselbergs. Point pattern analysis consists of a set of techniques which describes such spatial distributions in terms of the degree of clustering, randomness or regularity displayed with respect to an area defined by the investigator. In this sense, the term 'pattern' has a somewhat restricted meaning (Hudson and Fowler 1972). Despite such restrictions, the techniques have proved useful, first, in simply *describing* point patterns and, secondly, in throwing light upon the *processes* that have produced them. The first part of this chapter discusses how point patterns with various characteristics may be generated, emphasising those produced by an independent random spatial process. After this, two techniques are described, one based on **quadrat analysis**, which requires a count of points in each cell of a regular grid placed over a study area; the other upon **nearest neighbour analysis**, which relies on direct measurements of the spacing between points.

RANDOM AND SYSTEMATIC POINT PATTERNS

Suppose we wish to imitate or 'model' the process of locating 50 points randomly and independently within a region. This may be done with respect to a hypothetical square region such as that shown in Figure 8.1a. A square grid is placed over the region, each set of grid lines — one representing 'eastings' (U), the other 'northings' (V) — being numbered 00 to 99, starting from an origin in the south-west corner of the region in each case. It is much easier to use a random number table (Table A.7) than a physical representation such as the urn model. Remembering to enter this table at a randomly chosen point, read off two pairs of numbers. The first pair represents the easting of the first point to be located, the second pair its northing. If the first two pairs of numbers drawn are 28 and 43, the coordinates of the point $U_1 = 28$ and $V_1 = 43$ or (28, 43) (Fig. 8.1a). This procedure is repeated until the desired number of points has been located. Because each digit in a random number table is generated independently and randomly, it implies that any two pairs of two-digit numbers are also

(a) Random point pattern

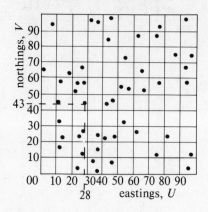

(b) Pattern generated by a deterministic process

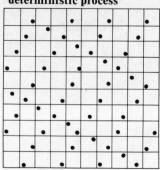

(c) Pattern generated by process with clustering tendencies

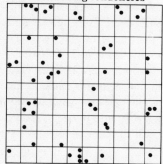

(d) Pattern generated by process with dispersion or regularity tendencies

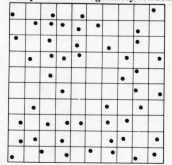

(e) Pattern generated by process operating at two spatial scales

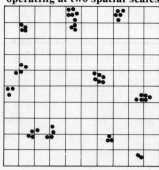

Figure 8.1   Examples of spatial point patterns.

generated independently and randomly. The 50 points in Figure 8.1a may therefore be regarded as a sample realisation of an independent random process operating in two-dimensional space.

Because there are 10 000 points to choose from in the range of coordinates from (00, 00) to (99, 99), the probability of any given point being the first to be chosen is 1/10 000 or 0·0001, i.e.

$$P(U_1 = I, V_1 = J) = 0·0001 \qquad (I, J = 0, 99)$$

No matter how many points have been selected, the probability of any point in the range (00, 00) to (99, 99) being chosen at any given stage remains 0·0001, if we assume sampling with replacement, so that

$$P(U_i = I, V_i = J \mid [U, V]) = P(U_i = I, V_i = J) \qquad (8.1)$$

$$(I, J = 0, 99, i = 1, 2, \dots)$$

where (U_i, V_i) is the ith point chosen, $[U, V]$ represents the set of coordinates of the $(i - 1)$ points already chosen, and $P(U_i = I, V_i = J) = 0·0001$. Equation 8.1 represents the statistical definition of independence, this time expressed in terms of a process generating points in two-dimensional space.

The set of 50 points in Figure 8.1a is usually described as a **random point pattern**, this being taken as shorthand for a sample realisation of an independent random spatial process. Slightly different patterns may be produced by subsequent realisations of 50 points.

In contrast to these irregular patterns, produced by a purely stochastic process, point patterns which appear to be perfectly ordered have almost certainly been produced by a deterministic process in which the 'events' (points) are completely dependent upon one another. An example of such a pattern is shown in Figure 8.1b. A great variety of completely deterministic, systematic or dependent spatial point patterns may be constructed using lattice theory (Coxeter 1961).

Most real-world 'spatial series' look as though they have been produced by a process which has both deterministic and stochastic components. A point pattern that is more regular (or more uniform) than random may be produced if each point is generated randomly, and a rule devised which makes it more likely that places far away from already located points will receive points than places nearby. The point pattern in Figure 8.1d was obtained in this manner, by ensuring that none of the 100 cells or 'quadrats' into which the region is divided could receive more than one point. If any particular location has a greater chance of receiving a point because it lies beyond some critical distance, d^*, from a pre-existing point, then

$$P(U_i = I, V_i = J \mid d_i > d^*) > 0·0001 \qquad (8.2)$$

where d_i represents the distance between (U_i, V_i) and a pre-existing point.

Equation 8.2 shows that processes characterised by tendencies to regularity or dispersion are not spatially independent.

> Develop a similar argument to show that a more clustered than random pattern, as shown in Figure 8.1c, is generated by a non-independent spatial process.

A variety of processes leading to various degrees and forms of regularity or clustering may be envisaged. Compound processes may also be at work, e.g. shops or plants may tend to occur in clusters, but these clusters may in turn be randomly distributed over a region, as in Figure 8.1e.

THE POISSON PROBABILITY DISTRIBUTION AND QUADRAT ANALYSIS

A point pattern may be summarised in terms of the frequency distribution of the number of points in quadrats. A **quadrat** is the basic areal unit or 'cell' into which we choose to divide a study region. It may be of any shape or size (in theory at least), but is usually square or rectangular. All the regions in Figure 8.1 have been subdivided into 100 square quadrats. The frequency distribution of the number of points is obtained by counting the number of quadrats containing 0 points, the number containing 1 point, the number containing 2 points, and so on until we have accounted for all the quadrats. Table 8.1 gives the frequency distributions for each of the point patterns in Figure 8.1. Compared with the independent random process, notice how the regular deterministic and more regular than random processes give results showing fewer empty quadrats, many more quadrats with one point, and fewer quadrats with two or more points. The results of simple and compound clustering processes show a different tendency, with rather more empty cells and cells containing two or more points than a pattern produced by a random process.

As was the case with the stochastic processes considered in Chapters 5 and 6,

Table 8.1   Frequency distributions of points per quadrat for point patterns of Figure 8.1.

Process	Row	Points						
		0	1	2	3	4	5	6
Independent random	(1)	58	35	6	1	0	0	0
Regular deterministic	(2)	51	48	1	0	0	0	0
Clustering random	(3)	69	18	9	2	2	0	0
More regular than random	(4)	50	50	0	0	0	0	0
Compound	(5)	80	6	5	5	2	1	1
Theoretical Poisson	(6)	60·65	30·33	7·58	1·26	0·16	0·02	

an independent random spatial process may produce a slightly different result each time. Some realisations may, quite by chance, appear to be 'more regular than random', others to be 'more clustered than random'.

> Use an independent random process to generate 50 points in a square region similar to those shown in Figure 8.1. Compare the resulting pattern visually with that of Figure 8.1a. Then, construct the frequency distribution of points per quadrat. Does the pattern 'happen' to be slightly more regular, or slightly more clustered, than that shown in Figure 8.1a?

If a variety of patterns may be generated by the same independent random spatial process, how then do we define a 'random point pattern'? A convenient yardstick is provided by the **Poisson probability distribution**, which gives the long-run proportions of quadrats containing given numbers of points, assuming that the process generating the points is independent and random.

We can imagine the individual probabilities of the Poisson distribution being derived in the following way. Suppose that the square in Figure 8.2 represents any *individual* quadrat taken from a square region such as Figure 8.1, and that the quadrat is in turn subdivided into a 'very large' number, r, of very small square cells. Because the process we are considering is independent and random, the array of r square cells can be treated just as if they were placed end to end to form a one-dimensional spatial series. If the probability of any given cell 'receiving' a point is p, then the probability of each of k cells in the quadrat

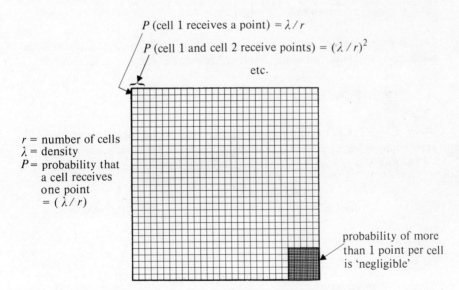

P (cell 1 receives a point) $= \lambda / r$

P (cell 1 and cell 2 receive points) $= (\lambda / r)^2$

etc.

r = number of cells
λ = density
P = probability that a cell receives one point $= (\lambda / r)$

probability of more than 1 point per cell is 'negligible'

Figure 8.2   Subdivision of individual quadrat to obtain Poisson probabilities.

receiving a point is given by the binomial probability

$$P(X = k) = \binom{r}{k} p^k q^{r-k} \tag{8.3}$$

where X is the number of cells receiving a point, i.e. the number of points in a quadrat. If there were $r = 100\,000$ cells in a quadrat, and the probability p was 'small', say $p = 0\cdot01$, then the (binomial) probability that a quadrat contains 3 points would be

$$P(X = 3) = \binom{100\,000}{3} (0\cdot01)^3 (0\cdot99)^{99\,997}$$

and might set us a few problems in calculation. However, a very close approximation may be obtained if it is possible to obtain the *density* (or average number of points per quadrat. This may sometimes be derived theoretically, but is more often estimated from a set of observations. In the case of the random point pattern shown in Figure 8.1a, the number of points is $n = 50$, the number of quadrats is $m = 100$, so that

$$\lambda = n/m = 50/100 = 0\cdot5$$

where λ (lambda) is an estimate of the density of points per quadrat. If an individual quadrat contains r cells, the probability of any given cell receiving a point is $p = (\lambda/r)$, and the probability of any given cell not receiving a point is $q = [1 - (\lambda/r)]$. Hence

$$P(X = k) = \binom{r}{k} (\lambda/r)^k [1 - (\lambda/r)]^{r-k}$$

Letting r become 'infinitely large', so that every cell becomes 'infinitesimally small' and capable of containing at most one point, it can be shown that

$$P(X = k) = \binom{r}{k} (\lambda/r)^k [1 - (\lambda/r)]^{r-k} \approx e^{-\lambda} \lambda^k / k! \tag{8.4}$$

where $e = 2\cdot71828\ldots$ and represents the base of the natural logarithms. The expression on the right-hand side of Equation 8.4 is a **Poisson probability**.

The probability that any given quadrat in Figure 8.1a contains no points is

$$P(X = 0) = \frac{(2\cdot71828)^{-0\cdot5}(0\cdot5)^0}{0!} = \frac{1}{(2\cdot71828)^{0\cdot5}} = 0\cdot6065$$

and

$$P(X = 1) = \frac{(2\cdot71828)^{-0\cdot5}(0\cdot5)^1}{1!} = \frac{0\cdot5}{(2\cdot71828)^{0\cdot5}} = 0\cdot3033$$

Making use of the relationship $P(X = k) = P(X = k - 1) \times (\lambda/k)$ (which holds only for Poisson probabilities), we obtain

$$P(X = 2) = 0.3033 \times (0.5/2) = 0.0758$$

$$P(X = 3) = 0.0758 \times (0.5/3) = 0.0126$$

$$P(X = 4) = 0.0126 \times (0.5/4) = 0.0016$$

The sum of the probabilities calculated so far is 0.9998. Because the remaining individual probabilities are so small, they are combined, so that by subtraction

$$P(X \geqslant 5) = 1 - 0.9998 = 0.0002$$

The long-run **expected numbers** of quadrats containing given numbers of points are obtained by multiplying the appropriate probability by the total number of quadrats, m, i.e.

$$E(X = k) = P(X = k) \times m \qquad (8.5)$$

where $E(X = k)$ is the expected number of quadrats containing k points. The expected numbers for the process and region we have been considering are given in the final row of Table 8.1, and have all the usual properties of expected values.

Check the expected numbers in Table 8.1, taking $m = 100$. Describe how you might set up an experiment designed to show that the theoretically derived expected numbers are closely approximated in practice.

A TEST FOR SPATIAL RANDOMNESS BASED ON THE VARIANCE—MEAN RATIO (χ^2 TEST)

This test exploits a distinctive property of the Poisson distribution — that its mean (μ) and variance (σ^2) are equal. It follows that the sample (or observed) mean or density (λ) and the sample (or observed) variance (s^2) of the number of points per quadrat for a pattern generated by an independent random process should be approximately equal — any difference will be due purely to random variation. The terms **sample mean** and **sample variance** are appropriate here because the spatial pattern from which they are calculated can be regarded as just one of the many possible realisations that might be produced by a stochastic spatial process.

Under the null hypothesis of randomness, it can be shown that the ratio of the (sample) variance and the (sample) mean or density, multiplied by the number of quadrats (m) minus one, follows the χ^2 distribution with $(m-1)$ degrees of freedom, i.e.

$$\chi^2_{m-1} = (m-1)(s^2/\lambda) \tag{8.6}$$

If the process giving rise to a point pattern incorporates a definite clustering or agglomerating component, the sample variance of the number of points per quadrat is liable to be increased. Compared with a random pattern, there will be more empty quadrats and more quadrats containing a relatively large number of points, so the spread about the mean is greater. By contrast a process characterised by a definite component leading to regularity or dispersion will produce relatively few empty quadrats, and relatively few quadrats containing many points, hence reducing the variance. The variance–mean ratios (VMR's) for the clustered, random and dispersed point patterns shown in Figure 8.1 were:

Clustering process	$s^2 = 0.838$ $\lambda = 0.500$	VMR = 1.676
Random process	$s^2 = 0.434$ $\lambda = 0.500$	VMR = 0.868
Dispersion process	$s^2 = 0.253$ $\lambda = 0.500$	VMR = 0.506

> Do the VMR values appear to be consistent with the arguments we have just mentioned?

Example

The technique has been applied to the spatial distribution of retail outlets by Guy (1976). In order to study the spatial distribution of Post-Offices, he placed a set of quadrats formed by a network of kilometre grid squares over the area. A quadrat was included in the analysis only if at least half its area was built-up or it contained at least one Post-Office (Fig. 8.3). The number of Post-Offices was $n = 51$, and the number of quadrats $m = 54$, giving $\lambda = n/m = 0.94$. To estimate the variance of the hypothesised Poisson process, use the computational formula for grouped frequency data (p. 20) which gives a precise answer in this case (why?). The formula should be modified, dividing through by $(m-1)$ rather than m, to ensure that σ^2 is not underestimated. It can be shown that the expected value of s^2 is $E(s^2) = \sigma^2$, as long as we use the formula given below:

$$s^2 = \frac{1}{m-1}\left[\sum_i f_i X_i^2 - \left(\sum f_i X_i\right)^2 \Big/ m\right] \tag{8.7}$$

EXAMPLE 103

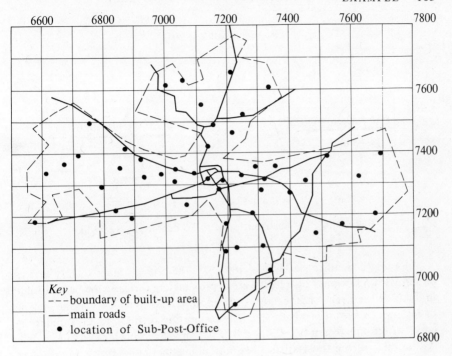

Figure 8.3 Location of Sub-Post-Offices in Reading.

The frequency distribution of Sub-Post-Offices in quadrats was found to be

Number of points	(X_i)	0	1	2	3
Frequency	(f_i)	15	30	6	3

so that

$$s^2 = (1/53)\{[(15 \times 0^2) + (30 \times 1^2) + (6 \times 2^2) + (3 \times 3^2)] - (51)^2/54\} = 0\cdot619$$

and

$$VMR = 0\cdot619/0\cdot94 = 0\cdot659$$
$$\chi^2_{53} = 53 \times 0\cdot659 = 34\cdot93$$

Unless independent evidence on the nature of the processes giving rise to the point pattern is available, a *two-tailed* test is appropriate, because the extreme values reflecting clustering and dispersion occur at opposite ends of the χ^2 distribution. The statistical hypotheses are

H_0: $\sigma^2 = \mu$ (the pattern is random, i.e. generated by an
independent random process),
H_1: $\sigma^2 \neq \mu$ (the pattern is non-random).

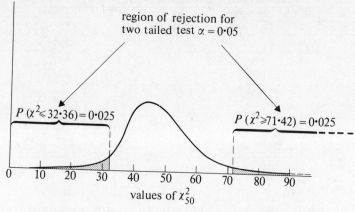

Figure 8.4 χ^2 distribution with 50 degrees of freedom.

Adopting a significance level of $\alpha = 0.05$, the critical values for a χ^2 distribution with 53 degrees of freedom are approximately (because the nearest value in Table A.1 is given for 50 degrees of freedom) 32·36 and 71·42, and the region of rejection is shown in Figure 8.4. The proportion of the total area under the curve within this region is 0·05 or 5%, 0·025 or 2·5% being located in each tail. As the value of the test statistic does not fall in either tail defining the region of rejection, we are unable to reject H_0 at the 5% significance level. For the number of degrees of freedom employed here, the χ^2 distribution is only slightly right-skewed; for df ≥ 100, the normal approximation may be used (see Q. 4, p. 111).

A *one-tailed* test may be carried out if evidence is available, from a source other than the point pattern itself, that the process giving rise to the point pattern was characterised by a systematic component.

If the component is thought to cause clustering, a one-tailed test based on the statistical hypotheses

$$H_0 : \sigma^2 = \mu \text{ (the pattern is random)},$$
$$H_1 : \sigma^2 > \mu \text{ (a clustering process is at work)},$$

is carried out. For a test at the $\alpha = 0.05$ significance level, the region of rejection lies above 67·50, as we expect a VMR > 1.0, and hence a χ^2 value greater than its mean of 53·0.

> State the statistical hypotheses that are appropriate, using both words and symbols, if independent evidence suggests a process leading to regularity or dispersion. Identify the region of rejection corresponding to (i) 5% significance level, (ii) 1% significance level.

Post Office policy in Britain is not to permit Post-Offices to be located too close together, so their spatial pattern should tend to show a systematic dispersion over an area. Significance testing produces a borderline result, since the critical lower tail value is 34·76, and the observed value is 34·93. We should reject H_0 because of prior knowledge of the processes operating (Post Office policy) as well as an 'almost significant' value of χ^2.

Advantages and limitations of quadrat analysis

The test described above readily allows us to compare an observed point pattern with that generated under the null hypothesis of an independent random spatial process, and to specify a directional alternative hypothesis (one-tailed test) if independent evidence is available. The quadrats used to divide up the study area may be of any shape, at least in theory, and ideally they should be small relative to the overall size of the study area, so that a large number of quadrats is obtained. There should be at least 50 to 100 quadrats, and preferably rather more — over 500 quadrats have been used in some studies, e.g. Getis (1964). However, quadrats should not be too small either — one rule of thumb suggested is that quadrat size should approximately equal twice the mean area per point. It is also possible to compare observed frequency distributions such as those in rows (1) — (5) of Table 8.1 with a variety of expected distributions generated by different processes under the null hypothesis. In the simplest case, the expected frequency distribution would be that generated by the Poisson distribution (row (6) in Table 8.1), but expected frequencies based on clustering or dispersion processes may also be derived. A useful introduction to the comparison of observed and expected frequency distributions of points using χ^2 or

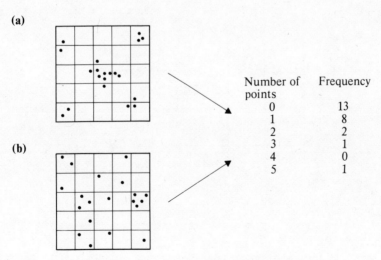

Number of points	Frequency
0	13
1	8
2	2
3	1
4	0
5	1

Figure 8.5   Relationships between frequency distributions and point patterns.

Kolmogorov–Smirnov tests of significance may be found in Taylor (1977, pp. 143–6, 169–70).

One of the most important limitations is that the results are strongly influenced by the number and size of the quadrats employed. Holding quadrat size constant, if the area included in the analysis is increased, so is the number of quadrats. If area is held constant, the number of quadrats may be increased by placing a grid with a finer mesh over the study area. In either case, the density λ will be altered, sometimes radically. This in turn alters the Poisson probabilities and may change the observed frequency distribution and the observed variance. The boundary of the study area must be defined with great care, and it has sometimes been suggested that quadrat analysis should be carried out at a number of different scales (i.e. quadrat sizes).

Finally, differences in spatial point patterns which are obvious to the naked eye may not be distinguished by quadrat analysis. The two patterns in Figure 8.5 would give rise to the same frequency distribution of points per quadrat.

NEAREST-NEIGHBOUR ANALYSIS

This technique avoids some of the problems associated with quadrat analysis and can in a sense be regarded as more truly 'geographic' in that it makes use of distances measured in two-dimensional space. Characteristics of point patterns are measured on a 'scale' ranging from 'perfectly clustered' to 'perfectly dispersed'.

The technique is based on a comparison of the *observed* mean first-nearest-

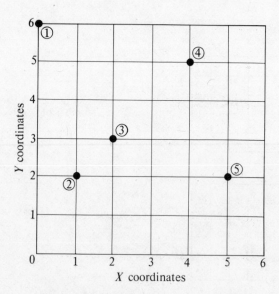

Figure 8.6   Locations of points used in hypothetical example.

neighbour distance, $\bar{d}$, calculated with respect to an actual point pattern, and the *expected* mean first-nearest-neighbour distance, δ^*, derived under the assumption that the point pattern was generated by a spatially independent random process. A simple hypothetical example first shows how the computations are carried out (Fig. 8.6).

To calculate the distance d_{ij} between any pair of points (U_i, V_i) and (U_j, V_j), apply the formula based on Pythagoras' theorem

$$d_{ij} = \sqrt{(U_i - U_j)^2 + (V_i - V_j)^2} \qquad (8.8)$$

Distances between points in Figure 8.6 are given in Table 8.2, and the minimum values in each row are italicised and recorded in the right-hand margin. The observed mean first-nearest-neighbour distance is given by

$$\bar{d} = \frac{\Sigma_i \min d_{ij}}{n} \qquad i = 1, \ldots, n \qquad (8.9)$$

where n is the total number of points, and for the hypothetical pattern in Figure 8.6, $n = 5$, $\bar{d} = 2\cdot49$. This value of $\bar{d}$ may be compared with the mean first-nearest-neighbour distance, δ^*, expected under the hypothesis that the observed number of points has been located independently and randomly in the study area, where

$$\delta^* = 1/(2\sqrt{\lambda}) \qquad (8.10)$$

and λ is the estimated density of points *per unit area*, i.e. $\lambda = n/A$, where A is the size of the study area. It is extremely important that the observed distances and the study area should be measured in the same metric — if the area is in square kilometres, distances must be in kilometres, if the area is in square miles, distances must be in miles etc. For the hypothetical region of Figure 8.6, $\delta^* = 1/(2\sqrt{5/36}) = 1/0\cdot746 = 1\cdot34$, assuming each square to be of side 1 km.

Table 8.2   Interpoint distances for hypothetical point pattern of Figure 8.6.

From point (i)	To point (j)					Minimum distance (min d_{ij})
	1	2	3	4	5	
1	0·00	4·12	*3·61*	4·12	5·39	3·61
2	4·12	0·00	*1·41*	4·24	4·00	1·41
3	3·61	*1·41*	0·00	2·83	3·17	1·41
4	4·12	4·24	*2·83*	0·00	3·17	2·83
5	5·39	4·00	*3·17*	3·17	0·00	3·17
					Total	12·43

$$\bar{d} = \frac{12\cdot43}{5} = 2\cdot49.$$

The nearest-neighbour index, R

If points are randomly scattered over an area, the ratio

$$R = \bar{d}/\delta* \qquad (8.11)$$

should be close to 1·0. If the points were produced by a clustering process, we should expect $\bar{d} < \delta$, so that $R < 1·0$, and if produced by a dispersion process, we expect $\bar{d} > \delta$, so that $R > 1·0$. R is known as the **nearest-neighbour index**, and is often used simply to describe point patterns without carrying out any test of significance. The theoretical limiting value of R is 0·0 for a perfectly clustered pattern (all points piled on top of one another) and 2·149 for a perfectly dispersed or regular pattern, the latter being defined as a set of points arranged in the form of a triangular lattice (Clark and Evans 1954) (Fig. 8.1b), and representing the basic settlement pattern envisaged by Christaller in his formulation of Central Place Theory.

The nearest-neighbour indices calculated with respect to a number of retail and service activities in Reading are shown in Table 8.3, ranked in descending order of their degree of dispersion. The highest value of $R = 1·589$ ($\bar{d} = 0·818$, $\delta = 0·515$) for the spatial pattern of Post-Offices confirms the result previously obtained in quadrat analysis.

If a pattern of points is generated by an independent random process, and the number of points, n, is relatively large — at least 50, and preferably 100 or more — the sampling distribution of $\bar{d}$ is normal with mean $\delta*$ ($= 1/(2\sqrt{\lambda})$) and standard error (the standard deviation of a sampling distribution is, by convention, called a **standard error**)

$$\sigma_{\bar{d}} = \tfrac{1}{2}\sqrt{(4/\pi) - 1} = \frac{0·26136}{\sqrt{n\lambda}} \qquad (8.12)$$

Table 8.3   Nearest-neighbour indices for selected retail and service activities in Reading.

Service	Nearest-neighbour index (R)	Number of outlets
Sub-Post-Offices	1·589	51
Fresh fish	1·207	20
Launderette	1·171	35
Funeral services	1·076	5
Newsagents	1·063	109
Men's hairdressing	1·051	41
Women's hairdressing	1·036	79
Medicines (dispensary)	1·029	34
Greetings cards	0·964	118
Dry cleaning	0·913	32
Women's outer clothing	0·492	100
Antique furniture	0·199	19
Building societies	0·090	27

The test statistic is

$$z = \frac{\bar{d} - \delta^*}{\sigma_{\bar{d}}}$$

(8.13)

One- or two-tailed tests may be carried out. For the Sub-Post-Office data, a one-tailed test of the form

$$H_0: \delta = \delta^* \text{ (the pattern is random)},$$
$$H_1: \delta > \delta^* \text{ (the pattern is dispersed)},$$

is appropriate. It is assumed that $\bar{d}$ provides an estimate of δ. Since $n = 51$, $\lambda = 0 \cdot 94$, we have

$$\sigma_{\bar{d}} = \frac{0 \cdot 26136}{\sqrt{51 \times 0 \cdot 94}} = 0 \cdot 038$$

Guy found that $\bar{d} = 0 \cdot 818$, $\delta^* = 0 \cdot 515$, and so

$$z = \frac{0 \cdot 818 - 0 \cdot 515}{0 \cdot 038} = 7 \cdot 97$$

which is highly significant, since $z_{0 \cdot 01} = 2 \cdot 33$ (one-tailed).

The answer is rather more emphatic than that given earlier by the χ^2 test on the variance–mean ratio.

Advantages and limitations of nearest-neighbour analysis

Nearest-neighbour analysis has the advantage of not being affected by the arbitrary choice of quadrat size, and of providing a scale of values against which to 'measure' the score of any observed point pattern. The technique is extremely sensitive to the way in which the study area is bounded. If a boundary is placed as shown in Figure 8.7a, the point pattern appears clustered, and the value of R

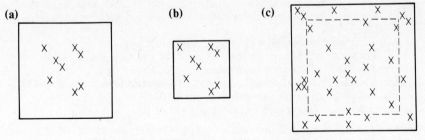

Figure 8.7   Influence of boundary effects.

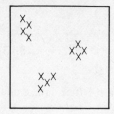 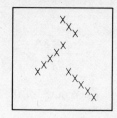 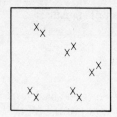

Figure 8.8   Clustered patterns indistinguishable using first-order nearest neighbours.

decreases because λ, and hence δ, will be deflated, compared with a situation in which the boundary is more tightly drawn (Fig. 8.7b). Boundaries should be drawn in a manner consistent with the subject matter under investigation (this also applies to quadrat analysis) – if the points represent retail outlets, the study area boundary should include all contiguous built-up areas at the edge of a city, and not simply follow administrative boundaries. Ideally, boundaries should be placed *within* the outermost limits of the study area as shown in Figure 8.7c. Otherwise, points located near the boundary are liable to have, on average, rather fewer neighbours than those nearer the centre of the area, and the observed mean first-nearest-neighbour distance, and the value of R, will tend to be biased upwards. Implausible results may be obtained because the technique, as so far described, cannot for example distinguish between the patterns shown in Figure 8.8, describing them all as 'clustered'. It is possible to include mean second, third or even higher-order nearest neighbours, and overcome such problems (Miller and Kahn 1962, Ch. 16). It is only possible statistically to test an observed mean first-nearest-neighbour distance against a null hypothesis of randomness, so in this sense the technique lacks the versatility of quadrat analysis. Finally, the probability distribution of $\bar{d}$ will be skewed, rather than normal, unless the number of points is at least 50, and preferably 100.

EXERCISES

1. For each of the phenomena described, state whether you think they are *statistically* independent or dependent over space. Explain each answer very briefly.
   (a) Locations of local shopping centres within a city.
   (b) Locations of settlements near a scarp foot and springlines.
   (c) Locations of solution hollows within a limestone region.
   (d) Locations of solution hollows within the British Isles.
   (e) Locations of major car dealers in a city.
   (f) Raindrops falling on an open area of downland.

2. Guy (1976) found that there were in all $n = 100$ retail outlets selling women's outer clothing in Reading. Placing $m = 53$ quadrats over the area, the frequency distribution of points per quadrat was found to be as shown below.

Number of points	0	1	2	3	4	5	...	60
Number of quadrats	33	11	1	3	2	2	...	1

(a) Determine the expected frequency distribution of points in quadrats assuming a completely random pattern of 100 points over the area.

(b) Calculate the variance—mean ratio with respect to the observations. Does its value agree with that obtained from the comparison of frequency distributions?

3. Smalley and Unwin (1968) identified individual drumlins from $1\cdot25\ 000$ topographic maps in three areas in County Clare, Ireland, and one in the Vale of Eden, England. Results of nearest-neighbour analysis are given below.

Location	Size of study area (km²)	Density (No./km²)	Number	$\bar{d}$	R
Co. Clare	24·08	1·33	32	0·564	1·31
Co. Clare	24·08	1·53	37	0·554	1·38
Co. Clare	56·16	1·62	91	0·498	1·28
Vale of Eden	40·04	0·92	37	0·584	1·13

(a) What kind of spatial pattern do the results suggest for each area?

(b) Can a statistical test be carried out with respect to any of these results? If so, carry it/them out, and justify your action. Comment on the result.

4. Getis (1964) carried out a study of changes in grocery store locations in Lansing, Michigan, over the period 1900 to 1960. He hypothesised that the pattern would reflect the influence of the following factors.

(i) Before major town growth in 1900—10, the pattern reflects varying population densities in the town at that time.

(ii) As the twentieth century progressed the very uneven population distribution would lead to clustering as small groups of stores served local concentrations of population.

(iii) In the most recent period, this tendency should reverse because car ownership permits an evening-out of population densities and reorganisation of the trade into fewer and larger outlets serving a more mobile population.

Year	Frequency (points per quadrat)						n	m	VMR	R	z
	0	1	2	3	4	5					
1900	89	13	3	0	0	0	19	105	1·13	1·074	+0·63
1910	112	20	4	3	0	0	37	139	1·44	0·673	−3·59
1920	261	43	12	2	3	1	110	322	1·31	0·658	−6·35
1930	315	66	18	5	3	0	129	407	1·47	0·772	−4·93
1940	389	82	15	5	2	2	145	495	1·56	0·792	−4·59
1950	433	76	15	5	1	0	125	530	1·34	0·841	−3·30
1960	598	64	4	0	0	0	72	657	1·00	0·998	−0·04

n = number of points, m = number of quadrats, VMR = variance—mean ratio, R = nearest-neighbour index, $z = (\bar{d} - \delta)/\sigma_{\bar{d}}$, where $\bar{d}$ is observed first-nearest-neighbour distance, and δ is expected first-nearest-neighbour distance.

(a) Do the results, overall, appear to support Getis' hypothesised pattern of change? (Do *not* carry out any statistical tests at this stage.)

(b) Carry out appropriate tests of significance on each of the variance—mean ratios and nearest-neighbour indices (i.e. values of R). Think carefully about the form the alternative hypothesis should take. (For degrees of freedom $\nu > 100$, use the approximation to the χ^2 distribution given by

$$z = \sqrt{2\chi^2} - \sqrt{(2\nu - 1)}$$

where z is a standard normal deviate.)

(c) What precautions should Getis have taken to ensure valid results?

REFERENCES AND FURTHER READING

General problems associated with the measurement of 'pattern' are discussed in:

Hudson, J. C. and P. M. Fowler 1972. The concept of pattern in geography. In *Man, space and environment*, P. W. English and R. C. Mayfield (eds), 545—50. Oxford: Oxford University Press.

Rogers, A. 1969. Quadrat analysis of urban dispersion: 1. Theoretical techniques. *Environ. Plann.* 1, 47—9.

For examples of use of the variance-mean ratio, see:

Getis, A. 1964. Temporal analysis of land use patterns with nearest neighbour and quadrat methods. *Ann. Assoc. Am. Geogs* 54, 391—9.

Guy, C. M. 1976. *The location of shops in the Reading area*. Reading: University of Reading, Dept of Geography, Geographical Paper no. 46.

A broader discussion of quadrat analysis is given in:

Harvey, D. W. 1966. Geographical processes and the analysis of point patterns: testing models of diffusion by quadrat sampling. *Trans Inst. Br. Geogs* 40, 81—95.

Taylor, P. J. 1977. *Quantitative methods in geography: an introduction to spatial analysis*. Boston: Houghton Mifflin. (pp. 134—150).

Thomas, R. W. 1977. *An introduction to quadrat analysis*. Concepts and Techniques in Modern Geography (CATMOG), no. 12. Norwich: Geo. Abstracts.

Nearest-neighbour analysis has been more widely employed. The seminal article is:

Clark, P. J. and F. C. Evans 1954. Distance to nearest neighbour as a measure of spatial relationships in populations. *Ecology* 35, 445—53.

For comments on advantages and disadvantages at an introductory level, see:

Pinder, D. A. and M. E. Witherick 1972. The principles, practice and pitfalls of nearest-neighbour analysis. *Geography* 57(4), 277—88.

———, ——— 1973. Nearest-neighbour analysis of linear point patterns. *Tijdschr. Econ. Soc. Geog.* 64(3), 160—3.

See, also, the critical article:

Dawson, A. H. 1975. Are geographers indulging in a landscape lottery? *Area* 7(1), 42—5; and four short articles in response in (1976) *Area* 8(3), 161—71.

An application to the study of settlement patterns is:

King, L. J. 1962. A quantitative expression of the pattern of urban settlements in selected areas of the United States. *Tijdschr. Econ. Soc. Geog.* 53, 1—7; also in *Spatial analysis*, (1968) B. J. L. Berry and D. F. Marble (eds), pp. 159—67.

For studies at the intra-urban scale, see Getis (1964) and Guy (1976) referenced above.

In physical geography, examples are:

Faniran, A. 1974. Nearest-neighbour analysis of inter-inselberg distance: a case study of the inselbergs of southwest Nigeria. *Z. Geomorphol.* (Supplement) 20, 150–67.

McConnell, H. and J. M. Horn 1972. Probabilities of surface Karst. In *Spatial analysis in geomorphology*, R. J. Chorley (ed.), pp. 111–34. London: Methuen.

Smalley, I. J. and D. J. Unwin 1968. The formation and shape of drumlins and their distribution and orientation in drumlin fields. *J. Glaciol.* 7, 377–90.

For treatment of higher-order nearest neighbours, see Taylor (1977, pp. 163–7) as referenced above, and:

Miller, R. L. and J. S. Kahn 1962. *Statistical analysis in geological sciences.* New York: Wiley. (Ch. 16).

Lattice theory is treated by:

Coxeter, H. S. M. 1961. *Introduction to geometry*. New York: Wiley.

9 Relationships Over Space II: Area Patterns

INTRODUCTION

We may wish to deal not only with maps depicting various phenomena as patterns of points but also with maps on which a score or measurement is associated with each point or, more commonly, with choropleth maps on which areas are subdivided into a set of contiguous (i.e. adjoining) regions. Such regions are often defined administratively or politically, as in the case of the boroughs of Greater London, states of the USA or countries of Africa, but may also be defined specially for a study, e.g. physiographic regions, quadrats laid out for a biogeographical investigation. The scores studied might be standard mortality ratios for a given disease in the counties of England, persons per household for census tracts in a North American city, or the number of different plant species observed in each quadrat on an area of heathland.

The study of spatial patterns of scores over a set of point locations or contiguous regions forms part of statistical theory concerned with **spatial auto-correlation**. If any systematic spatial pattern can be identified in the scores, those scores are to be spatially autocorrelated. There are two major reasons why the study of spatial autocorrelation has assumed an important role in geography. First, it represents a logical extension of the one-dimensional runs test (Ch. 6) to the task of distinguishing systematic from random variation in two-dimensional spatial patterns. Secondly, many of the inferential techniques described in Section III should be used only if the values obtained are spatially independent which, in the terminology of this chapter, implies that they should show no spatial autocorrelation, i.e. no systematic variation in space. Thus, the first task of spatial autocorrelation studies represents a direct contribution to the search for order in spatial series, while the second is a means for testing whether one of the assumptions underlying the use of various statistical procedures is satisfied. The second issue is discussed where appropriate in Section III.

PRELIMINARY DEFINITIONS AND CONCEPTS

Although it is possible to deal with measurements obtained at the ordinal scale or above, and with nominal variables of more than two categories, the basic ideas can be grasped if attention is confined to the *binary* case, i.e. nominal variables

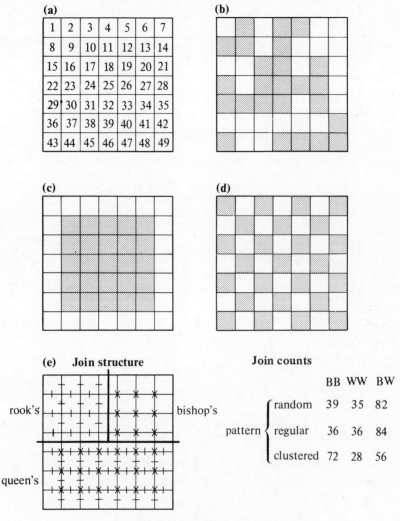

Figure 9.1   Regular lattice patterns.

with only two categories. We also deal only with scores over a set of contiguous regions, as these form the spatial framework for the vast majority of spatial autocorrelation studies.

Conventionally, each region in the study area may be coloured black (B) or white (W), the colours representing values of the binary variable. Black regions may be those in which a particular activity or quality is present, or in which the value of a variable measured at a higher level lies above a value such as the mean or median. The regional framework may consist of a **regular lattice** (Fig. 9.1a) or, as is more often the case in human geography, of an **irregular lattice** (Fig. 9.2a), of areas for which data are available (numbers in regions represent a

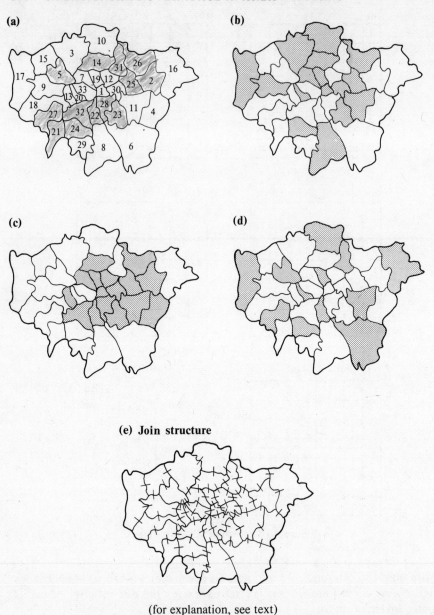

(a)

(b)

(c)

(d)

(e) Join structure

(for explanation, see text)

Figure 9.2   Irregular lattice patterns.

convenient labelling system *not* scores). The pattern of scores may appear to have been arranged randomly (Fig. 9.1b and 9.2b), the result of deterministic processes leading to perfectly clustered or perfectly regular (alternating) patterns (Fig. 9.1c and 9.1d) or produced by processes representing a mixture of systematic and unsystematic (i.e. random) forces (Fig. 9.2c and 9.2d). Where no particular spatial pattern is apparent, there is said to be *no* spatial autocorrelation. If adjacent regions tend to have highly similar scores, this is known as *positive* spatial autocorrelation; if scores in adjacent regions tend to be dissimilar, this is known as *negative* spatial autocorrelation. Of course, many spatial patterns other than those given in Figures 9.1 and 9.2 may be imagined.

> Label each of the maps in Figures 9.1 and 9.2 according to the kind of spatial autocorrelation it apparently displays. Can you think of any real-world patterns of scores which show (a) positive, (b) negative, spatial autocorrelation?

SPATIAL AUTOCORRELATION ON REGULAR LATTICES

Measures of spatial autocorrelation can be obtained by counting the number and kind of **joins** between regions. The join structure for the regular lattice of Figure 9.1 is shown in Figure 9.1e. Joins are defined where regions meet at a vertex (point) as well as along an edge (line) but, as shown later, the definition of a join may depend upon what we wish to find out. For the moment all possible joins are considered, as indicated by the *queen's case* (so-called by analogy with chess moves). Joins are described as BB, BW or WW if they connect two black regions, a black and a white region, or two white regions, respectively.

Compared with the pattern in Figure 9.1b, in which 25 black and 24 white scores were allocated randomly to the 49 regions, the number of BB and WW joins observed in a pattern produced by systematic forces leading to clustering tends to be relatively large, and the number of BW joins relatively small. This is confirmed by the join counts recorded at the foot of Figure 9.1. Intuitively, we expect a relatively large number of BW joins, and correspondingly reduced numbers of BB and WW joins for a pattern produced by systematic forces leading to regularity of the kind shown in Figure 9.1d. However, the table at the foot of Figure 9.1 shows virtually no difference between the join counts for this pattern and the random case, a point to which we return later.

CONTIGUITY TEST FOR SPATIAL AUTOCORRELATION

This test, based on the join counts described above, enables us to decide how likely it is that any given pattern of black and white scores was produced by a purely random process, and may be carried out under either of two sets of assumptions: those of free sampling or those of non-free sampling.

The test may be carried out under the assumption of **free sampling** only if the probability that a region will be black or white can be determined without reference to the study area itself, i.e. the probability is known *a priori*. For instance, the probability of any sub-region of England and Wales (an irregular lattice) having a standard mortality rate at or above the average could be calculated from the national figures, and this probability applied to a constituent area such as that consisting of the London boroughs. However, it is difficult to think of a case in physical geography where the probability of a given quadrat in a regular lattice being black can be determined in this way. The probability of a region being black in free sampling will be denoted by p, and of a region being white by $q = (1 - p)$.

A pattern produced under the free-sampling assumption represents sampling *with replacement* from an appropriate urn model. Consider the 49 regions in Figure 9.1, and take $p = 0\cdot5$. We place ten balls in an urn, five of which are black, and five of which are white, representing black and white regions respectively. The order in which scores are assigned to regions does not affect the long-term pattern of results, so the labelling system of Figure 9.1a is used. If a white ball is drawn first, region 1 is coloured white. The ball is replaced, and a second drawing made. This time the ball happened to be black. The pattern in Figure 9.1b resulted from 49 such drawings.

> How would you imitate this urn model, using random number tables?

Repeating the entire procedure many times with respect to the 7×7 regular lattice of Figure 9.1, approximations to the sampling distributions of BB, BW and WW joins may be obtained. If certain assumptions are fulfilled, then the sampling distributions of the join counts are known to be approximately normal. Taking $p = q = 0\cdot5$, Figure 9.3a shows sampling distributions of BB, BW and WW joins derived by generating 200 maps by computer on a 7×7 regular lattice, and counting the number of such joins for each map.

The *assumptions* required to ensure approximate normality are as follows (Cliff and Ord 1973, Ch. 2):

(1) the number of regions, n, in the study area should be greater than 20, and preferably 30 or more, and
(2) neither p nor q should be less than $0\cdot2$.

There are a number of other useful guidelines. Experimental work has shown that approximations to normality are better for values of p or q greater than $0\cdot5$, e.g. $0\cdot7$ or $0\cdot8$, than they are for such values lying an equivalent 'distance' below $0\cdot5$, e.g. $0\cdot3$ or $0\cdot2$. Thus, if $p = 0\cdot3$, $q = 0\cdot7$, the join count for WW rather than BB joins should be examined. If any one region or set of regions is joined to a large proportion of all other regions in the study area, as is the central region in Figure 9.4, then the normality assumption is unlikely to hold. Dacey (1968) also remarks that if the study area is highly elongated, such as Chile or peninsular Italy, the number of regions should be 100 or more.

If satisfied that our data conform to these guidelines, the **contiguity test** may be used. Formulae for the mean or expected numbers of joins, and their variances, for the free-sampling case are given in Table 9.1a.

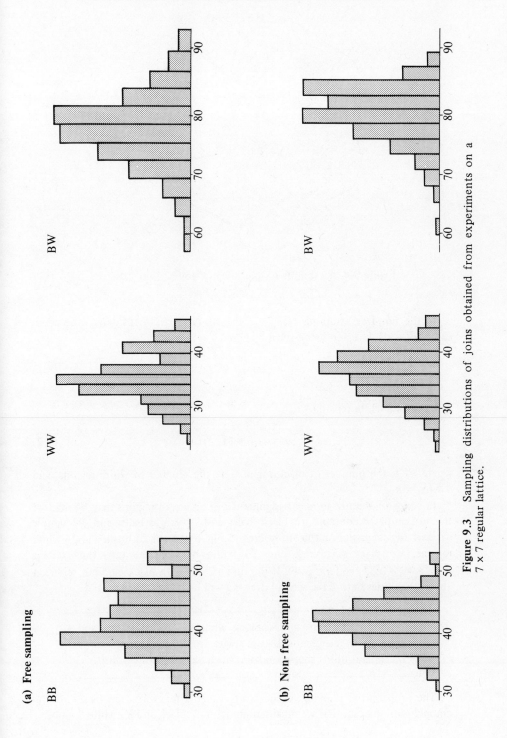

Figure 9.3   Sampling distributions of joins obtained from experiments on a 7 × 7 regular lattice.

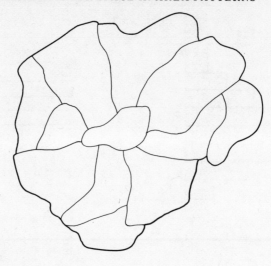

Figure 9.4   Example of 'badly conditioned' join structure.

In the **non-free sampling** case only the study area is considered, the probability of a black region being given by

$$p = n_1/n \tag{9.1}$$

and that of a white region by

$$q = n_2/n = (1 - p) \tag{9.2}$$

where n_1 is the number of black regions, n_2 the number of white regions, and $n = (n_1 + n_2)$.

In terms of the urn model, assignment of colours to regions may be made if the urn initially contains n_1 black balls and n_2 white balls, and we sample *without replacement*. As the study area in Figure 9.1 has 25 black and 24 white regions, we place 25 black and 24 white balls in the urn, thus setting $p = 25/49 = 0.51$, and $q = 0.49$. If the first ball drawn is black, and not replaced, the probabilities before the second drawing will be $p = q = 24/48 = 0.5$.

Assuming the first ball was black, what will the values of p and q be (i) if the second ball is white, (ii) if the second ball is black, and (iii) if the second and third balls are both white?

The probabilities p and q are therefore altered after every drawing in order to comply with the fact that we must end up with n_1 black and n_2 white regions.

Table 9.1   Formulae for expected values and variances for the contiguity test.

(a) Free sampling.

Means (expected values)	$\mu(BB) = Jp^2$ $\mu(BW) = 2Jpq$ $\mu(WW) = Jq^2$	p = probability that a region is black q = probability that a region is white L_i = number of joins associated with ith region. $J = \frac{1}{2}\Sigma L_i$ $K = \frac{1}{2}\Sigma L_i(L_i - 1)$
Variances	$\sigma^2(BB) = \mu(BB) + 2Kp^3 - (J + 2K)p^4$ $\sigma^2(BW) = 2(J + K)pq - 4(J + 2K)p^2q^2$ $\sigma^2(WW) = \mu(WW) + 2Kq^3 - (J + 2K)q^4$	

(b) Non-free sampling.

Means (expected values)

$$\mu(BB) = \frac{Jn_1(n_1 - 1)}{n(n - 1)} \qquad \text{for calculation of } J \text{ and } K, \text{ see above.}$$

$$\mu(BW) = \frac{2Jn_1 n_2}{n(n - 1)}$$

$$\mu(WW) = \frac{Jn_2(n_2 - 1)}{n(n - 1)}$$

Variances

$$\sigma^2(BB) = \mu(BB) + \frac{2Kn_1(n_1 - 1)(n_1 - 2)}{n(n - 1)(n - 2)}$$

$$+ \frac{[J(J - 1) - 2K]n_1(n_1 - 1)(n_1 - 2)(n_1 - 3)}{n(n - 1)(n - 2)(n - 3)}$$

$$- [\mu(BB)]^2$$

$$\sigma^2(BW) = \mu(BW) + \frac{4[J(J - 1) - 2K]n_1(n_1 - 1)n_2(n_2 - 1)}{n(n - 1)(n - 2)(n - 3)}$$

$$+ \frac{2Kn_1 n_2}{n(n - 1)} - [\mu(BW)]^2$$

$$\sigma^2(WW) = \mu(WW) + \frac{2Kn_2(n_2 - 1)(n_2 - 2)}{n(n - 1)(n - 2)}$$

$$+ \frac{[J(J - 1) - 2K]n_2(n_2 - 1)(n_2 - 2)(n_2 - 3)}{n(n - 1)(n - 2)(n - 3)}$$

$$- [\mu(WW)]^2$$

Sampling distributions of BB, BW and WW join counts in the non-free sampling case are shown in Figure 9.3b, again based on 200 maps generated by computer for a 7 x 7 regular lattice.

> The variety of spatial patterns of scores that could possibly be produced on the same lattice is very much greater under the assumption of free sampling. Why should this be so? How is this feature reflected by the variances of the join counts shown in Figure 9.3?

Notice that it is possible to determine means and variances of BB, BW and WW join counts for any kind of regular (or irregular) lattice, either by carrying out a large number of sampling experiments or, more readily, by using the formulae given in Table 9.1.

Statistical hypotheses and test statistics

Assuming no prior information about the nature of the patterning in the regional scores, the statistical hypotheses are:

H_0: the spatial pattern of scores is random, i.e. has been produced by a random process,

H_1: the spatial pattern of scores is non-random, i.e. has been produced by a non-random process.

Because of the assumed normality of the sampling distributions of join counts, significance tests are based on the standard normal deviate, z, whose critical values are tabulated in Table A.3. We calculate the **standard errors** of the join counts and the test statistics (z values) from the formulae given in Table 9.1.

Since

$$J = J(BB) + J(BW) + J(WW)$$

i.e. the total number of observed joins, J, is given by adding together the observed numbers of BB, BW and WW joins, only two of the join counts contain independent information. Tests should be performed on the BW count and on only one of the BB or WW counts. Choice of BB or WW should be in accordance with the previously stated guidelines. If $z(BW)$ and the other z value are both significant at the desired level, it seems reasonable to reject H_0.

The statistical hypotheses can be stated more specifically in terms of join counts. If $\mu^*(BW)$ is the (unknown) expected number of BW joins produced by the processes at work in the study area, and $\mu(BW)$ the expected number of BW joins produced under the null hypothesis of no spatial autocorrelation, then we have

H_0: $\mu^*(BW) = \mu(BW)$,
H_1: $\mu^*(BW) \neq \mu(BW)$.

> Define statistical hypotheses in terms of join counts for BB and WW joins.

Examples

Let us see how well the z test performs with respect to the three patterns shown in Figures 9.1b, 9.1c and 9.1d. The tests will be carried out under the assump-

tion of non-free sampling, as this is generally more realistic. No prior information on the nature of the processes at work is available, so that non-directional
hypotheses (two-tailed) for each pattern are as given at the end of the previous
section.

> Would you choose the BB or WW join counts for inclusion in
> the test in this case? Is the argument in favour of your choice
> particularly strong?

The two constants J and K required for the calculations are, from Table 9.2,

$$J = \tfrac{1}{2} \sum_{i=1}^{n} L_i = \frac{312}{2} = 156$$

(note the division by 2 because summing the L_i counts each join twice), and

$$K = \tfrac{1}{2} \sum_{i=1}^{n} L_i(L_i - 1) = \frac{1824}{2} = 912$$

and $n_1 = 25, n_2 = 24, n = 49$. The expected numbers of joins are

$$\mu(BB) = \frac{156 \times 25 \times 24}{49 \times 48} = 39 \cdot 80$$

$$\mu(BW) = \frac{2 \times 156 \times 25 \times 24}{49 \times 48} = 79 \cdot 59$$

Table 9.2   Calculation of L and K for the pattern of Figure 9.1b.

Region	Colour	BB	WW	BW	L_i	$L_i(L_i - 1)$
		Type of join				
1	W	0	0	3	3	3 x 2 = 6
2	B	2	0	3	5	5 x 4 = 20
3	W	0	1	4	5	5 x 4 = 20
4	B	2	0	3	5	5 x 4 = 20
⋮	⋮	⋮	⋮	⋮	⋮	⋮
19	W	0	2	6	8	8 x 7 = 56
⋮	⋮	⋮	⋮	⋮	⋮	⋮
47	B	2	0	3	5	5 x 4 = 20
48	B	3	0	2	5	5 x 4 = 20
49	B	2	0	1	3	3 x 2 = 6
	Totals	78	70	164	312	1824

$J(BB) = 39$	$J(WW) = 35$	$J(BW) = 82$	$J = 156$
$K = 912$	$n_1 = 25$	$n_2 = 24$	$n = 49.$

with variances and standard errors given by

$$\sigma^2(\text{BB}) = 39 \cdot 80 + \frac{2 \times 912 \times 25 \times 24 \times 23}{49 \times 48 \times 47}$$

$$+ \frac{[(156 \times 155) - (2 \times 912)] \times 25 \times 24 \times 23 \times 22}{49 \times 48 \times 47 \times 46} - (39 \cdot 80)^2$$

$$= 39 \cdot 80 + 227 \cdot 70 + 1334 \cdot 76 - 1584 \cdot 04 = 18 \cdot 22$$

and so

$$\sigma(\text{BB}) = 4 \cdot 27$$

and

$$\sigma^2(\text{BW}) = 79 \cdot 59 + \frac{4 \times [(156 \times 155) - (2 \times 912)] \times 25 \times 24 \times 24 \times 23}{49 \times 48 \times 47 \times 46}$$

$$+ \frac{2 \times 912 \times 25 \times 24}{49 \times 48} - (79 \cdot 59)^2$$

$$= 79 \cdot 59 + 5824 \cdot 40 + 465 \cdot 31 - 6334 \cdot 57 = 34 \cdot 73$$

and so

$$\sigma(\text{BW}) = 5 \cdot 89$$

Reference to the foot of Figure 9.1 shows that the observed numbers of joins for the randomly generated pattern are $J(\text{BB}) = 39$, $J(\text{BW}) = 82$, and so

$$z(\text{BB}) = \frac{39 - 39 \cdot 80}{4 \cdot 27} = -0 \cdot 19$$

$$z(\text{BW}) = \frac{82 - 79 \cdot 59}{5 \cdot 89} = 0 \cdot 41$$

The signs of these test statistics may be ignored because the test is two-tailed. Reference to Table A.3 shows that neither z value exceeds the critical value of $1 \cdot 96$ for the 5% significance level, so we feel unable to reject H_0.

> Does the test of significance lead us to make the correct decision in this case?

The expected values and standard errors are the same for the patterns of scores shown in Figures 9.1c and 9.1d, because the values of J, K, n_1, n_2 and n

remain unaltered. However, the numbers of BB, BW and WW joins are different.

Count the number of BB, BW and WW joins for the patterns in Figures 9.1c and 9.1d. Then, confirm that the test statistics for the *clustered pattern* are

z(BB) = 7·54 z(BW) = –4·01 z(WW) = –2·04

and for the *regular pattern* are

z(BB) = –0·89 z(B W) = 0·75 z (WW) = –0·14

On the basis of these values, what do you conclude about the nature of the processes giving rise to the spatial patterns of scores in each case? Are any of your conclusions surprising? If so, in what way?

SEARCHING FOR SPATIAL AUTOCORRELATION IN PARTICULAR DIRECTIONS

You should have noticed that the z scores calculated for the perfectly alternating pattern of Figure 9.1d would make us extremely reluctant to reject the null hypothesis of randomness. This counter-intuitive result occurs because we have been searching for spatial autocorrelation in all possible directions on the lattice, taking into account the joins between adjacent regions in rows (east–west) (north-west–south-east and north-east–south-west). As already mentioned, this search is called the **queen's case**. Retaining the chess analogy, the north–south and east–west searches are known as the **rook's case**, and those along the diagonals as the **bishop's case**. The relevant join structures are illustrated in Figure 9.1e.

If the perfectly alternating pattern is examined according to the rook's case alone, or the bishop's case alone, the join counts of Table 9.3 are recorded.

Using the expected values and standard errors already given for the non-free sampling case, determine the values of z for the rook's and bishop's cases. Do the results seem to accord with your intuition this time?

Does it make any difference if the analysis is repeated under the free sampling assumption, taking $p = q = 0·5$?

Table 9.3 and Figure 9.1e also show that the queen's case may be regarded as the addition, or superimposition, of the rook's and bishop's cases, leading to a

Table 9.3   Join counts for a 7 x 7 regular lattice showing a perfectly alternating spatial pattern — rook's, bishop's and queen's cases.

	K	J	J(BB)	J(BW)	J(WW)
Rook's case	214	84	0	84	0
Bishop's case	170	72	36	0	36
Queen's case	912	156	36	84	36

cancelling of the basic patterns of the latter two cases and interpretation of the *overall* pattern as random.

As Cliff and Ord emphasise, the investigator must decide whether to search for spatial autocorrelation which has no directional bias, as in the queen's case, or to detect patterning in specific directions. For example, a physical geographer may suspect that geological structures which show a north—south trend influence the spatial pattern of surface features, or that a particular vegetation type tends to occur on superficial deposits which show a south-west—north-east trend across a study area. Evidence of stronger clustering in a north—south than an east—west direction should be expected in the first case, and in a south-west—north-east rather than a north-west—south-east direction in the latter case. (Notice also that z values based on half of the joins in the rook's case (north—south) are compared with z values based on the remainder (east—west) for the first example, and a similar strategy adopted for the second.) Cliff (1968) used the rook's case to compare north—south and east—west trends in the spatial growth pattern of numbers of adopters of an innovation in Sweden, as the physical grain of the country, and the orientation of roads and railways, suggested a north—south spread of information.

There seem to have been virtually no applications of the simpler spatial autocorrelation techniques described here to test for trends in particular directions in space, but there are surely many opportunities to be seized.

Such applications also depend upon appreciation of the additional information that may be provided by one-tailed tests.

ONE-TAILED TESTS

If prior information is available about the processes giving rise to the observed pattern of scores, then a *one-tailed* test is appropriate. For instance, if we are led to believe that a clustering process is at work, the hypotheses are

H_0: the pattern is random,
H_1: the pattern is clustered.

Statement of hypotheses in terms of join counts is more complex than usual, because the direction (or tail) of the test depends upon the type of join examined. In this case, we should expect 'too many' BB joins, 'too few' BW joins, and 'too many' WW joins, and set up the hypotheses

$$H_0: \mu^*(BB) = \mu(BB), \qquad H_0: \mu^*(BW) = \mu(BW), \qquad H_0: \mu^*(WW) = \mu(WW),$$
$$H_1: \mu^*(BB) > \mu(BB), \qquad H_1: \mu^*(BW) < \mu(BW), \qquad H_1: \mu^*(WW) > \mu(WW),$$

where the $\mu(\cdot)$ represent expected values under the null hypothesis of randomness.

> How would you modify H_1 in each case to deal with a situation in which processes were thought to lead to regularity or dispersion?

Remember that the critical values of z for a one-tailed test differ from the corresponding two-tailed values.

SPATIAL AUTOCORRELATION ON IRREGULAR LATTICES

The formulae presented in Table 9.1 for the free and non-free sampling cases may be applied without modification to irregular lattices of the kind shown in Figure 9.2. This particular lattice represents the 33 Greater London boroughs, and Figure 9.2b gives the spatial pattern of mortality rates for 1974 (see *Annual Abstract of Greater London Statistics 1974* 1976). Boroughs with rates above the median for all Greater London boroughs are coloured black, the remainder white. The mortality rates have been corrected for differences in factors such as sex, age and marital status. The number and type of joins associated with each borough is given in Table 9.4.

> Should the assumption of free or non-free sampling be adopted in this case? Are the assumptions and guidelines underlying the use of the contiguity test satisfied?

All the basic information required to carry out the contiguity test is given in Table 9.4. You are also given that

$$\sigma^2(BB) = 7 \cdot 98 \qquad \sigma^2(BW) = 17 \cdot 93 \qquad \sigma^2(WW) = 8 \cdot 25$$

> Carry out the contiguity test with respect to the pattern shown in Figure 9.2c, carefully justifying your choice of sampling assumption and join counts.

Table 9.4 Joins for the London boroughs data.

Borough	Colour	Type of join BB	WW	BW	L_i	$L_i(L_i-1)$
1 City	B	6	1	0	7	7 × 6 = 42
2 Barking	B	4	1	0	5	5 × 4 = 20
3 Barnet	W	0	2	3	5	5 × 4 = 20
4 Bexley	B	2	2	0	4	4 × 3 = 12
5 Brent	W	0	2	5	7	42
6 Bromley	W	0	5	1	6	30
7 Camden	B	3	3	0	6	30
8 Croyden	W	0	1	3	4	12
9 Ealing	W	0	1	4	5	20
10 Enfield	W	0	1	2	3	6
11 Greenwich	B	5	1	0	6	30
12 Hackney	B	5	1	0	6	30
13 Hammersmith	B	1	5	0	6	30
14 Haringey	B	5	1	0	6	30
15 Harrow	W	0	0	4	4	30
16 Havering	W	0	3	0	3	6
17 Hillingdon	W	0	0	3	3	6
18 Hounslow	W	0	1	3	4	12
19 Islington	B	4	0	0	4	12
20 Kensington	W	0	2	2	4	12
21 Kingston	W	0	1	3	4	12
22 Lambeth	B	3	4	0	7	42
23 Lewisham	B	3	1	0	4	12
24 Merton	W	0	2	3	5	20
25 Newham	B	5	1	0	6	30
26 Redbridge	B	2	2	0	4	12
27 Richmond	W	0	2	2	4	12
28 Southwark	B	4	2	0	6	30
29 Sutton	W	0	0	3	3	66
30 Tower Hamlets	B	6	0	0	6	30
31 Waltham Forest	W	0	4	1	5	20
32 Wandsworth	B	2	5	0	7	42
33 Westminster	W	0	5	2	7	42
Totals		60	62	44	166	724

$J(BB) = 30$ $J(WW) = 22$ $J(BW) = 31$ $J = 83$

$K = 362$ $n_1 = 16$ $n_2 = 17$ $n_3 = 33.$

The distinction between rook's, bishop's and queen's cases cannot normally be made for irregular lattices because no consistent orientations can be assigned to contiguous regions.

ADVANTAGES AND LIMITATIONS OF THE CONTIGUITY TEST

The version of the test outlined above has the advantage of relative simplicity, requiring only binary data and join counts between regions. The main disadvantage is that much information is lost because of the low level of measurement achieved. A more complex procedure, which can be applied to nominal data in k categories – the k-colour test – is also available (Dacey 1968, Cliff and Ord

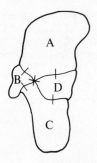 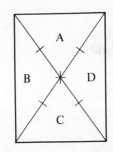 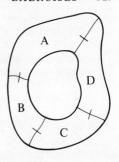

Figure 9.5   Three regional systems with the same join structure.

1973, pp. 6–7). Tests originally devised by Geary and Moran may be used for ordinal, interval or ratio data. Cliff and Ord (1973, pp. 7–10) provide a theoretical treatment, and formulae and examples are given in Ebdon (1977, pp. 133–7) and Haggett *et al* (1977, Ch. 11).

There are also limitations which the test shares with others for detecting spatial autocorrelation.

(a) The 'topological invariance problem'. Once joins have been defined, the size and shape of the regions, and the relative strength of the links between them (in terms of common boundary length, road and rail connections, political links etc) are ignored. For example, all three systems of regions in Figure 9.5 have the same join structure.

(b) Only joins between adjacent areas are considered. More detailed discrimination between patterns may be possible if joins between second-, third- or even higher-order nearest-neighbouring areas are considered (note the parallel with nearest-neighbour analysis of point patterns).

Extensions that allow for weighting of joins and treatment of higher-order joins are considered in Cliff and Ord (1973, pp. 11–13).

It is worth emphasising here that the translation of information about a spatial pattern provided by a map to a non-spatial representation – whether it be a frequency distribution, contingency table or a mean first-nearest-neighbour distance – always involves loss of information and a degree of distortion. Conversion of the perfectly alternating pattern of Figure 9.1d, for instance, to a set of join counts for the queen's case suppresses all evidence of this striking regularity. Such problems do not render these techniques invalid, but warn us to pay critical attention to what we are doing.

EXERCISES

1. Show that, under the free sampling assumption, the generation of black and white scores is independent and random. Is this also true in the case of non-free sampling? Explain your answer carefully.

2. London boroughs in which the infant mortality rate was above the average for England and Wales in 1974 are those numbered 2, 5, 14, 21, 22, 23, 24, 25, 26, 27, 28, 31 and 32 in Figure 9.2a and Table 9.4. Test this pattern of scores for evidence of spatial autocorrelation. Is the free or non-free sampling assumption appropriate?

REFERENCES AND FURTHER READING

An extended treatment of spatial autocorrelation is given in:
   Cliff, A. D. and J. K. Ord 1973. *Spatial autocorrelation.* London: Pion.
and summarised by
   Haggett, P., A. D. Cliff and A. E. Frey 1977. *Locational analysis in human geography*, 2nd edn. Vol. 2: *Locational methods*. London: Edward Arnold.
Another useful discussion is given in:
   Dacey, M. F. 1968. A review on measures of contiguity for two and *k*-colour maps. In *Spatial analysis*, B. J. L. Berry and D. F. Marble (eds), pp. 479–95. Englewood Cliffs: Prentice-Hall.
and a simple introduction in:
   Ebdon, D. 1977. *Statistics in geography: a practical approach*. Oxford: Blackwell. (pp. 128–41).
A more complex version of the contiguity test, the *k*-colour test, is described in Cliff and Ord (1973, pp. 6–7), Dacey (1968) and Ebdon (1977, pp. 133–7) as referenced above. Tests for ordinal, interval or ratio data are discussed by Cliff and Ord (1973, pp. 7–10), and Ebdon (1977, pp. 133–7) provides formulae and examples. Extensions which allow for weighting of joins and treatment of higher-order joins are considered in Cliff and Ord (1973, pp. 11–13).
   Reported applications of the test are still relatively few (for residuals about regression see Ch. 15); see, for example:
   Cliff, A. D. 1968. The neighbourhood effect in the diffusion of innovations. *Trans Inst. Br. Geogs* **44**, 75–84.
   Dalton, M. and J. M. Seaman 1973. The distribution of New Commonwealth immigrants in the London Borough of Ealing, 1961–6. *Trans Inst. Br. Geogs* **58**, 21–40. Ebdon (1977), as referenced above, also provides a number of examples.
The data for the example and exercise based on mortality rates are from:
   Annual abstract of Greater London statistics 1974 1976. Vol. 9. London: Greater London Council.

III

ARTIFICIAL SAMPLING CONTEXTS

10   Random Sampling Procedures

INTRODUCTION

In an artificial sampling context, we make use of a stochastic process with known properties, a **random sampling procedure**, in order to draw inferences about the characteristics of a set of all possible measurements on the basis of the information yielded by a subset or sample of those measurements. This contrasts with a natural sampling context in which we are trying to make inferences about the characteristics of the stochastic process which is at work.

POPULATIONS AND SAMPLES

The set of all possible measurements is known as a statistical **population**. The term should not be confused with its everyday usage meaning 'total number of people'. Strictly speaking, a statistical population is a population of **measurements**, but the term is used often to describe the entire set of people, places or other phenomena to which the measurements refer. Thus, we tend to say 'the population consists of all the pebbles in a glacial deposit' rather than 'the population consists of the pebble orientation measurements for all the pebbles in a glacial deposit'. The questions listed in Table 10.1 give examples of statistical populations that may interest geographers. The population with respect to Question 1 consists of all the replies (1 = yes, 0 = no) given by West Indian

Table 10.1   Some questions of interest to geographers.

1 Have the majority of West Indian households in Reading (England) migrated directly from the West Indies, rather than living elsewhere in Great Britain first?
2 Are the majority of households in the Reading area located within walking distance (1 km, say) of a shop selling bread?
3 Do valley-side slopes protected from recent basal cutting by a stream have lower angles than those being actively eroded?
4 Is the proportion of land with desirable soil properties in a region of an underdeveloped country at least 0·7 (70%), therefore justifying investment in an irrigation scheme?
5 What is the preferred orientation (if any) of the pebbles in a glacial deposit?
6 What is the spatial distribution of soil types in England and Wales?
7 How does mean annual rainfall vary over the British Isles?
8 What is the average travel time involved in the journey to work in London, and what proportions of individuals travel by train, bus, car, etc?
9 Does average annual household income vary from one area of the UK to another?
10 Do adults in two different areas of a town — one working-class, the other middle-class — differ in their degree of familiarity with photographs of town centre street scenes?

Table 10.2   Commonly encountered parameters and estimators.

| Parameter | | Estimator | | |
Description	Symbol	Description	Symbol	Formula
Number of observations or measurements	N	Number of observations	n	—
Population mean (interval or ratio measurements)	μ	Sample mean	$\bar{X}$	$\dfrac{1}{n}\Sigma X_i$
Population proportion (nominal measurements)	Π	Sample proportion	p	$m/n\dagger$
Population variance	σ^2	Sample variance	s^2	$\dfrac{1}{n-1}\Sigma(X_i-\bar{X})^2$
Population standard deviation	σ	Sample standard deviation	s	$\sqrt{\dfrac{1}{n-1}\Sigma(X_i-\bar{X})^2}$

† m denotes the number of measurements falling in the category or class of interest.

households in Reading. The population is 'large' ($N \simeq 2000$) but finite. To answer Question 4, information is required on the soil characteristics at every point on the map. Theoretically, a point is dimensionless (it has no area) so that the population is infinite. In practice, as samples would be taken at grid references, the population is 'very large', say 100 000 or more, depending upon the fineness of the coordinate system employed. Whenever the population of interest is too large for complete coverage, some means should be devised of obtaining a subset or sample of measurements.

Any summary measurement of the characteristics of a population is known as a **parameter**. The proportion of all West Indian households in Reading who migrated directly from the West Indies is therefore a parameter, as is the mean slope angle of all valley-side slopes in a study area. The corresponding quantity based on a sample of measurements is known as an **estimate** of that parameter, and the formula used to calculate the value of the estimate or statistic is called an **estimator**. Thus, the formula for the sample mean (Table 10.2) is an estimator which provides an estimate of the population mean.

Parameters are usually denoted by Greek symbols; estimators are sometimes denoted by lower case letters, sometimes by other conventions (see Ch. 15). Some frequently encountered parameters and their estimators are given in Table 10.2.

OBJECTIVES OF SAMPLING

There are two major statistical objectives of sampling. Some questions in Table 10.1 demand an estimate of a population parameter. To answer Question 1, we must find out what proportion of observations fall into the 'yes' category, and estimates of mean slope angles are needed to answer Question 3. The branch of statistical theory covering procedures designed to provide estimates is called **estimation**. Some questions in the table also involve **hypothesis testing**. In Question 1, we must decide whether at least 0·5 or 50% of the observations fall into the 'yes' category and, for Question 3, whether a difference between sample means reflects a real difference between the unknown population means. Hypothesis testing is based on **tests of significance**, already considered in a natural sampling context.

Because no sampling procedure is costless, and is also liable to produce errors, a further objective of sampling is *either* to provide the required information within tolerable limits of sampling error for the lowest cost *or*, for a fixed cost, to yield the required information with least sampling error.

The statistical objectives of sampling are of course subservient to the broader purpose of the investigation. A useful discussion of the aims of various social surveys is given in Moser and Kalton (1971, Ch. 1), and Beckett (1968) and Young (1973) provide interesting treatments of the relationship between the scale at which soil and land-use surveys are carried out and the purpose — reconnaissance, resource inventory etc — for which it is designed.

> For each question in Table 10.1, state whether you think estimation, hypothesis testing (or both) are required. Is any information needed which does not strictly fall under either heading?

ADVANTAGES AND LIMITATIONS OF SAMPLING

Some advantages are obvious. If the population of measurements is infinite, we have no other choice, and considerable amounts of time, money and effort would also be saved if the population is very large, say $N > 1000$. A further benefit is that more time and money can be devoted to ensuring the quality of the data, perhaps by training interviewers and field workers properly or employing professional agencies, and by instituting careful checks to reduce the number of errors liable to occur at various stages in the process from initial collection to final tabulation or statistical analysis (the Census of England and Wales has been criticised on these grounds by Gray & Gee (1972)). Results will also be available more rapidly from a survey of modest size, an important consideration to a planning department or retail company wanting, for example, information on changes in household travel and shopping behaviour after the opening of a large out-of-town supermarket. Sampling is desirable if the very act of obtaining measurements is 'reactive' or destructive. If all surveys were based on entire populations of towns or regions, individuals would soon become uncooperative because of being constantly pestered. Measurement of soil characteristics must be restricted to sample sites, partly because augering or the digging of pits is involved.

Under some circumstances, it is quite possible that estimates based on careful sample measurements may be as accurate, or even more so, than those based on a hurried census of the entire population; the greater reliability and validity are sufficient to offset sampling error.

The major limitation of sampling is **sampling error**, which represents the difference between a sample estimate and a population parameter. Unless certain precautions are taken, sample estimates may be **biased**, thus tending systematically to over- or underestimate the quantities of interest, or **imprecise**, meaning that the variation of the sample estimates around the population parameter is considerable, or both. To ensure minimum bias and maximum precision given the resources available, we must use a procedure based on **independent random sampling** (in American terminology this is known as **probability sampling**). If such procedures are used, the probable size of sampling error associated with an estimate can be determined.

Use of other methods for selecting items almost always introduces bias, often quite unconsciously. Bias may occur because it is tempting to include only certain areas or sites because they are more accessible, e.g. beaches and cliffs near a road, and certain individuals or households because they are at home at more convenient times of the day for interviewing. Non-random sampling need not produce valueless results – daily weather maps are based upon observations which are by no means located according to the principles of sampling theory. In such cases, however, estimates of sampling error cannot be made.

SAMPLING PROCEDURES

There are certain 'ground rules' to be observed when sampling, the two most important being that

(a) a sampling frame should be available which in some way lists all the entities (e.g. persons, places, regions) from which the sample is to be drawn, and
(b) measurements should be obtained with respect to entities which have been selected by the procedure of independent random sampling, or by some satisfactory approximation to it.

We emphasise that all sampling procedures differ in detail, because they should be tailored to the problem under consideration.

The **sampling frames** employed may actually list all the entities in question, or allow this in principle. If information on *individual* adults is required, as in Question 10 (Table 10.1), the British investigator is fortunate because the Register of Electors ('Electoral Register'), which is updated annually, lists every adult individual registered to vote in a ward or parish. For various reasons, some people fail to register, and it is possible that all those missing have rather similar characteristics in terms of what we are studying, thereby introducing bias. There is little we can do about this problem, however, and it is not regarded as being particularly serious. For the United States, there is no comparable list, but the US Census does report the number of housing units for every county and minor civil division over 1000 population (although this is done only every ten years), and for every block for all cities over 50 000 population. Backstrom and Hursh (1963) give a detailed account of a procedure for selecting households or individuals randomly, taking the sampling frame of city blocks and housing units as a starting point.

> For what kinds of individuals or households would a telephone directory provide a good sampling frame?

For sampling purposes, maps possess many of the properties of lists. For studies carried out within a small area, a large-scale map, e.g. the 1:1250 Ordnance Survey series, provides a **spatial sampling frame**, distinguishing and numbering individual dwelling units. A spatial sampling frame is necessary to estimate areal coverage (e.g. Question 4), the total set of items being represented by all the points on a map. Sampling frames of this type are particularly common in physical geography.

It is useful here to distinguish between the **target population**, and the **sampled population** (Krumbein and Graybill 1965, pp. 149–53). The target population represents the total set of measurements relevant to the study, e.g. all measurements of cobble orientation in a till deposit, or of stone roundness in a stream, but the sampled population represents measurements only for those areas or sites accessible to the investigator, such as glacial deposits exposed by fluvial or coastal erosion. Andrews (1971) points out that till fabrics are normally taken from the vertical face of an exposure so that all parts of the target population consisting of measurements for all pebbles in the till are not accessible. A further problem is that sampling may then be restricted to the lower and upper 3 m of, say, a 30 m vertical exposure. It becomes a matter of the geographer's judgement whether the sampled and target populations are sufficiently similar to allow inferences about the sampled population to be applied to the target population.

> What kinds of sampling frame are likely to be used for each question in Table 10.1? Do the frames yield sampled or target populations?

Independent random sampling and simple random sampling

Ideally, individuals or items should be selected from a sampling frame by independent random sampling. To answer Question 10 in Table 10.1, we might obtain the electoral registers for the area in question. Suppose there are $N = 3000$ names in the register, and a sample size of $n = 100$ individuals is desired. The names are numbered 0 to 2999 (making 3000 numbers in all) and then, entering a random number table at random as described on p. 67, draw four-digit numbers until one hundred within the range 0–2999 are obtained (any numbers exceeding 2999 would be discarded). If we sample *with replacement*, it is possible for the same name to be included in the sample more than once. This is an undesirable feature, for obvious reasons! However, sampling with replacement fulfils the assumption of *independent* random sampling. For this example, the probability of any given name being selected when the first random number is drawn is $p = 1/3000 = 0.00033$, and p remains constant for subsequent drawings if the name is replaced. Sampling *without replacement* alters p as each name is drawn.

> Show that the value of p just before the one-hundredth item is drawn differs by only 0·00001 from that at the beginning of the sampling process.

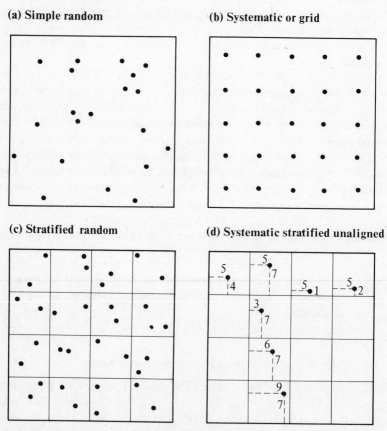

(a) Simple random   **(b) Systematic or grid**

(c) Stratified random   **(d) Systematic stratified unaligned**

Figure 10.1   Examples of point sampling designs (from Taylor 1977, p. 77).

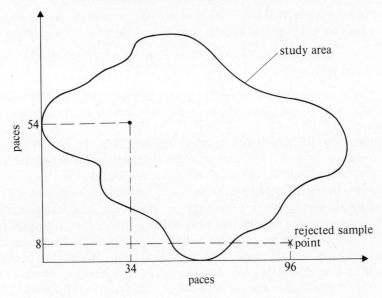

Figure 10.2   Random selection of points in the field.

In this case, and in most real-world problems, the absolute and relative differences between values of p are so small that they can be ignored. The procedure based on sampling without replacement described above is known as **simple random sampling**.

In order to locate sites by simple random sampling in the field for vegetation or soil sampling, the spatial equivalent of simple random sampling may be employed. If a map is available, points may be located randomly within the study area using random number tables in exactly the same way as described in Chapter 8 (p. 95) (Fig. 10.1a). Points falling outside an irregularly shaped study area may be ignored. This technique may also be used to estimate areal coverage, as in Question 4. The number of points falling within an area possessing the prescribed characteristics is simply expressed as a proportion of the total number of points in the sample. If no map is available, two axes at right angles may be established by field survey (Fig. 10.2). The desired number of sample sites may then be determined by first drawing a random number from Table 6.1 and taking that number of paces eastwards from the origin, and then drawing a second random number to give the number of paces to be taken northwards.

An important feature of a simple random sample is that it may not give even coverage of the sampling frame — the items selected may show signs of bunching in certain parts of the list or certain areas of a map. This is sometimes considered to be a disadvantage, particularly in the case of spatial sampling (Fig. 10.1a).

In general, simple random sampling is designed to give every item or person in the sampling frame a known and equal probability of inclusion in the sample and every possible sample of size n the same probability of being drawn. Sampling errors associated with simple random sampling are discussed in Chapter 11.

Systematic sampling

An alternative to simple random sampling is **systematic sampling** which is generally easier and quicker, and may give more precise results with a smaller sampling error.

Considering the hypothetical case in which $n = 20$ names are to be sampled from a list of $N = 100$, we first calculate $N/n = 5$ to obtain the sampling interval. The first name should be selected randomly from among the first N/n (= 5), and then every (N/n)th (i.e. fifth) name thereafter — hence the procedure is known as systematic sampling with a random start. This yields more even coverage of the sampling frame. Only five distinctly different systematic samples could be drawn from such a frame, compared with the $\binom{100}{20}$ (a number in thousands of millions) different simple random samples conceivably obtainable. However, a systematic sample of 100 items drawn from 10 000 provides 100 potentially different samples. Systematic sampling will give satisfactory results provided that the items in the sampling frame are more or less randomly arranged, at least in terms of characteristics which may influence the findings of our study. Otherwise, the sampling interval might coincide with some periodic feature in the sampling frame. For example, if all the streets in the survey area contain roughly the same number of inhabitants, and the interval (or some integer multiple of it) coincided with this number, then we might find that all the names selected are those of 'young singles' living in apartments near a main road, young marrieds with children living in semi-detatched houses back from the main road being left out! Most writers on sampling seem to agree that such periodicities are quite rare when sampling from lists, but in sampling from a map the problem is more severe.

'One-dimensional' systematic sampling has many applications in physical geography. Measurements of pebble size or shape, or valley-side slope angle may be determined at equal intervals along a narrow stream. Thus, the slope measurements necessary to answer Question 3 (Table 10.1) were obtained at intervals of 5 to 10 ft in fine-textured badlands or 100–150 ft in medium-textured topography (Strahler 1950, p. 678). The problem of spacing is crucial in studies of slopes, as the processes operating at one point along a valley side may well influence those operating on points very close by, and in this sense contravene the statistical assumption of independence. Choice of an interval depends very much upon the knowledge and judgement of the investigator.

'Grid sampling', or systematic sampling in two dimensions, is widely used and advocated in studies of soil and vegetation properties (Fig. 10.1b). Andrews (1971) has suggested that pebbles be taken from exposures of glacial tills at sites designated by a systematic grid (Fig. 10.3). Berry and Baker (1968) include it as one possible sampling procedure for estimating areal coverage factors, in the case

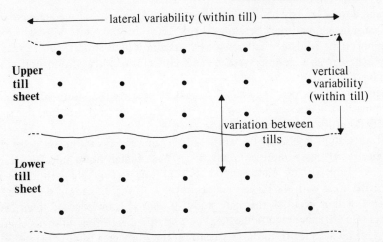

Figure 10.3   Systematic point sampling of till exposure (from Andrews 1971, p.8).

of land devoted to various uses in north-east Illinois. Ideally, the first point of the grid should be located randomly, and the remainder located at appropriate intervals, What is 'appropriate' depends on the sample size required and the precise aims of the study.

It is difficult to ascertain the magnitude of the sampling error associated with systematic samples, although it is almost always smaller than that for simple random samples, particularly for spatial sampling. However, the formulae for standard errors (sampling error) of a simple random sample are usually taken to apply, provided no discernible periodicities which may coincide with the sampling interval are present in the sampling frame.

Stratified sampling

Samples are **stratified** in order to make the maximum possible use of all the information the investigator has about the study area, to permit comparisons between different groups or areas, and to increase the precision with which estimates are made.

It is highly desirable (although not always possible) to stratify the population before sampling. Otherwise, in spatial sampling for many surveys in physical geography 'natural' differences between spatial units which influence the phenomenon under study, e.g. above and below a break of slope, may not be detected adequately. It is of little value to obtain four measurements within area A and 96 in area B when trying to estimate soil pH or amount of exchangeable cations in each! Andrews (1971) describes a hypothetical case in which two till sheets may be distinguished on a colour and lithological basis in the same exposure, as is shown schematically in Figure 10.3. A simple random or systematic sample of sites may be located within each unit to provide adequate coverage and permit comparisons between them. Each of the units into which the sampling frame is divided is known as a **stratum** (plural **strata**). Notice that the sample design also permits vertical and lateral variability within units to be examined.

The procedures described above can all be described as **stratified systematic sampling**. If a simple random sample were taken within each stratum, this would be called **stratified random sampling**.

Stratification is also useful when carrying out surveys in human geography, To ensure adequate representation of all West Indian households in Reading (Qu. 1, Table 10.1), Matowanyika (1977), an undergraduate student, first identified areas with the highest concentrations of West Indians in the town, and sampled within each of them. A study of adults' knowledge of town centre streets (Qu. 10) may require the area to be stratified according to socioeconomic status, or by distance from the city centre if the study area is fairly extensive.

However, it may be difficult to stratify in advance in human geography, except in a rather crude manner. For instance, data are not available on income and other socioeconomic data except for enumeration districts, or still larger units, so that selection of, for example, 'middle-class' and 'working-class' areas has to be based on rather subjective criteria — such as predominant housing type found in the areas — which allow much room for error. A study requiring distinctions between no-car, one-car and two-car households probably depends upon personal knowledge of an area and could involve much frustrated walking.

Students carrying out their own surveys often find it difficult to ensure an adequate sample size within each stratum of interest. This may be overcome by introducing a **quota** element into the procedure. If, say, 50 one-car and 50 two-car households are required, then a list of 150 or even 200 addresses or names should be obtained by simple random or systematic sampling. A record is kept of the number of addresses at which interviews have been obtained for each kind of household. If the quota of 50 one-car households is completed first, interviews are conducted at the remaining sample addresses on the list only if the households turn out to be two-car households. Many addresses will be rejected, hence the need for plenty of 'spares'.

It is always possible to stratify according to numerous factors which may affect the phenomenon under study, e.g. age, income, stage in life cycle. Unless a very large sample can be obtained (unlikely in the case of undergraduate thesis work) it is simply not possible to take all these factors into account, so you need to pick out factors you consider to be the most important, and stratify according to those.

> For which questions in Table 10.1 would stratification of the population be desirable? What problems do you foresee?

Sampling errors with respect to stratified procedures or designs are smaller than those associated with unstratified procedures. Calculation of such errors will not be considered in this text, and for details you should consult Gregory (1978, pp. 101–12), Moser and Kalton (1971, pp. 85–100) or Sampford (1962, Ch. 6).

The proportion of entities to be selected for inclusion in the sample from each stratum, known as the **sampling fraction** (f), need not be the same for each stratum. If a 10% ($f = 0.1$) sample is taken of one-car households in an area, for example, we may obtain 500 households, which is far too many to handle (in student thesis or project work). However, a 10% sample of two-car households may yield only 15 households, which is rather too few. Thus, a $2\frac{1}{2}\%$ ($f = 0.025$) sample of one-car households, and a 50% ($f = 0.5$) sample of two-car households may be far more appropriate. (For comments on sample size, see Ch. 11.)

Spatial stratified point-sampling designs may be useful in areal coverage studies and in physical geography. The 'cells' or 'quadrats' in Figures 10.1c and 10.1d represent the spatial strata. There are many possible designs, but the **spatial stratified random** and **spatial stratified systematic unaligned** designs are regarded as particularly useful (Berry and Baker 1968).

SAMPLING UNDER EXPERIMENTAL CONDITIONS

The assumption of independence may be fulfilled by experimental control. Suppose an investigator wished to know how much mineral matter (by weight) was eroded from a soil covered by scrub vegetation under conditions of heavy rainfall. Each of a number of bulk field samples may be subjected to a controlled rate of water flow for a fixed time period using an experimental flume. The weight of the material eroded from each of say, six or eight samples, can be regarded as a function of (i) a systematic component, representing the 'true' amount of material removed from such soils by the experimentally controlled flow rate, and (ii) a random component reflecting minor measurement errors and differences in the flow rate from one experiment to the next. The experimenter must ensure, for example, that the angle of the chute down which the water passes does not alter, or that sediment does not partly obstruct any portion of the system, as this would destroy the assumption of independence and impart bias. The mean of six or eight experimental results is taken as an estimate of the 'true' value. Such studies may be carried out in place of field studies on run-off plots, which are costly and time consuming.

'Quasi-experimental' studies by behavioural geographers may profitably employ **randomisation** procedures. When individuals are given long lists of rating scales or photographs, presentation of the list in a different random order to each individual ensures no systematic effect due to any particular ordering influences the results. Items presented on cards may be shuffled randomly between each interview; otherwise, only three or four different random orders should be used to avoid unnecessary typing.

OTHER DESIGN CONSIDERATIONS

Choice of individuals or sites in sampling constitutes only one aspect of the overall problem or research design. This is a very complex topic and just a few cautionary remarks can be made here.

It is extremely easy to *confound* the effects of two (or more) variables. If, when trying to answer Question 10 (Table 10.1), a sample of 100 middle-class individuals is chosen from a private estate on the southern edge of a city, and 100 working-class individuals from a public housing estate on the northern fringe of that city, then any differences in perception due to class factors will be indistinguishable from those due to the difference in area of residence. Similarly, if Strahler had been able to find valley-side slopes protected from basal cutting only on rock type A, and those being actively eroded only on rock type B, then there would be no way of disentangling the effects of erosional environment from those of rock type (Qu. 3, Table 10.1). Impact studies are also very prone to such difficulties. The impact of the opening of a new section of motorway is hard to gauge because an increase in employment, number of dwellings etc. may well have occurred anyway, perhaps due to a general increase in prosperity.

SCOPE OF INFERENCES BASED ON SAMPLE OBSERVATIONS

Statistical inferences based on sample observations apply only to the populations from which those samples were drawn. Findings obtained from a study of a *sample* of West Indian households in Reading can be extrapolated, bearing in mind the sampling error involved, to the population of West Indian households in Reading, *not* to all West Indians in south-east England or in Britain. Inferences made about slopes and slope processes from a sample of measurements in a study area in the Southern Uplands of Scotland apply only to that area. The results may seem capable of extrapolation to wider populations or areas, but this is very much a matter for the geographic knowledge and judgement of the investigator.

It is also important to remember that sampling error depends upon *sample size* rather than upon the *sampling fraction*, although a very high sampling fraction obviously implies a relatively small sampling error. A full discussion is given in Chapter 11.

EXERCISES

1. Comment on any problems encountered with the following sample designs.

   (a) One class of schoolchildren in a mixed-ability state school, another in a private school, are asked to rate each of the standard regions of England and Wales in terms of residential desirability. Half of the children in each class are provided with maps, and half not. The comprehensive school is mixed, the private school is a single-sex girls school.

(b) In a study of the journey to work, behaviour of middle-class households located on the fringe of a city was compared with that of working-class households located close to the city centre.

(c) Bulk soil samples were obtained from randomly chosen sites in an area which included a chalk escarpment, an outcrop of Upper Greensand and a zone of clay.

(d) A soil conservation agency wishes to find out which of two vegetation types (A or B) should be planted to reduce soil erosion most effectively. Which of the designs proposed below do you prefer?

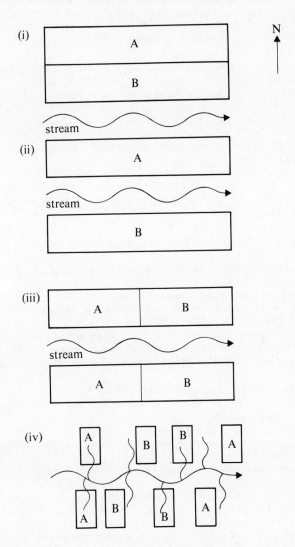

2. A geographer wishes to find out what people's attitudes towards different kinds of recreation area are like. Randomly selected groups were approached in a busy part of a heathland area used mainly by ramblers, near Birmingham. Comment on the meaning of the results obtained.

REFERENCES AND FURTHER READING

Theoretical and practical expositions, particularly with respect to survey research, are given in:

Backstrom, C. H. and G. D. Hursh 1963. *Survey research.* Evanston: Northwestern University Press.

Moser, C. A. and G. Kalton 1971. *Survey methods in social investigation*, 2nd edn. London: Heinemann.

Stuart, A. 1962. *Basic ideas of scientific sampling.* London: Griffin.

The above references also discuss research design, but for further information and examples see:

Kish, L. 1959. Some statistical problems in research design. *Am. Sociol. Rev.* **24**, 328–38.

Tanur, J. M. (ed.) 1972. *Statistics: a guide to the unknown.* San Francisco: Holden-Day.

General comments on soil and land-use surveys are given in:

Beckett, P. H. T. 1968. Method and scale of land-resource surveys, in relation to precision and cost. In *Land evaluation*, G. A. Stewart (ed.). London: Macmillan. pp. 52–63.

Young, A. 1973. Soil survey procedures in land development planning. *Geog. J.* **139**(1), 53–64.

For discussions of spatial sampling, see:

Berry, B. J. L. and A. Baker 1968. Geographic sampling. In *Spatial analysis*, B. J. L. Berry and D. F. Marble (eds), pp. 91–100. Englewood Cliffs (New Jersey): Prentice–Hall.

Haggett, P., A. D. Cliff and A. E. Frey 1977. *Locational analysis in human geography*, 2nd edn. Vol. 2: *Locational methods.* (pp. 267–82). London: Edward Arnold.

Sampford, M. R. 1962. *An introduction to sampling theory.* Edinburgh: Oliver and Boyd.

The concept of the target population is treated in:

Krumbein, W. G. and F. A. Graybill 1965. *An introduction to statistical models in geology.* New York: McGraw-Hill.

Stratified sampling and associated standard errors are discussed in Moser and Kalton (1971) and Sampford (1962) (see full references above), and in:

Gregory, S. 1978. *Statistical methods and the geographer*, 4th edn. London: Longmans.

The logic of randomisation techniques is explained by:

Edgington, E. S. 1966. Statistical inference and non-random samples. *Psychol. Bull.* **66**, 485–7.

Other examples referred to in the text are:

Andrews, J. T. 1971. *Techniques of till fabric analysis.* Br. Geomorphol. Res. Group Tech. Bull. no. 6.

Family expenditure survey, report for 1974 1975. London: H.M.S.O., Dept of Employment.

Gray, P. and F. A. Gee 1972. *A quality check on the ten per cent sample census of England and Wales.* London: H.M.S.O., Office of Population Censuses and Surveys.

Matowanyika, J. Z. Z. 1977. *The migration behaviour of West Indian immigrants to Reading.* University of Reading, Dept of Geography, Unpublished undergraduate dissertation.

Strahler, A. N. 1950. Equilibrium theory of erosional slopes approached by frequency distribution analysis. *Am. J. Sci.* **248**, 673–96, 800–14.

Two recent references on sampling and social survey methods, written specially for geographers, are:

Dixon, C. J. and B. Leach 1978. *Sampling methods for geographical research*. Concepts and Techniques in Modern Geography (CATMOG), no. 17. Norwich: Geo Abstracts.

Dixon, C. J. and B. Leach 1978. *Questionnaires and interviews in geographical research*. Concepts and Techniques in Modern Geography (CATMOG), no. 18. Norwich: Geo Abstracts.

11 Parametric Methods I: Theory and Estimation

INTRODUCTION

Assuming that information has been obtained according to the principles described in the preceding chapter, how can we make inferences about populations from samples? If **parametric methods** are employed, such inferences may be based on estimates, described in this chapter, or upon tests of hypotheses, considered in Chapter 12.

Parametric methods share a number of characteristics. Estimates are required of the parameters, in this case the means and variances, of the population frequency distributions from which sample measurements are obtained. The variance estimate is then used to derive the standard error of the quantity to be estimated, but in so doing it is necessary to make *distributional* assumptions about the form of the sampling distribution and, if the sample size is relatively small, about the frequency distribution of the population of measurements.

The first part of this chapter is theoretical, and deals with the sampling distribution of a proportion based on nominal measurements or count data, before generalising our findings by appeal to an important statistical theorem called the Central Limit Theorem. Application of these results to practical problems of estimation is then discussed, and we conclude by considering the relationship between sample size and sampling error.

DISTRIBUTION THEORY FOR A SAMPLE PROPORTION

The major problem of sampling in an artificial sampling context is that of **sampling error**. If the true proportion of adults residing in a small suburban estate who travel to work in the centre of the nearby metropolis is represented by Π, and the sample proportion by p, then

$$e_p = p - \Pi \tag{11.1}$$

where e_p represents sampling error. Let us consider the behaviour of such errors when simple random sampling is employed. Assume that a suitable sampling frame, whose form is outlined schematically in Table 11.1 is available, and that the true proportion is $\Pi = 0 \cdot 20$. Initially a sample size of $n = 10$ is used. The first sample yields an estimate of $p = 0 \cdot 4$, or a sampling error of

$$e_p = p - \Pi = 0 \cdot 4 - 0 \cdot 2 = +0 \cdot 2$$

Table 11.1   Schematic representation of a sampling frame.

	00	01	02	03	04	05	06	07	08	09	10	11	12	13	14	15	16	17	18	19
000																				
020																				
040																				
060																				
080																				
100																				
120																				
140																				
160																				
180																				
200																				
220																				
240																				
260																				
280																				
300																				
320																				
340																				
360																				
380																				

Key

▨ adult who works in centre of metropolis

☐ adult who does not work in centre of metropolis

so that p overestimates Π in this case.

> Use random number tables to draw a simple random sample of size $n = 10$ from the sampling frame in Table 11.1. Does p over- or underestimate Π?

Taking 100 samples of size $n = 10$ in a class experiment, the frequency distribution of sample proportions shown as in Figure 11.1a was obtained. This distribution is known as the **sampling distribution of a proportion**.

Because one of the highly desirable properties of any estimator, such as p, is that it should be unbiased, this implies the average of all the estimates yielded by the estimator, i.e. its **expected value**, should equal the value of the parameter being estimated. Although the sampling distribution in Figure 11.1a is strongly right-skewed, the average of the 100 sample proportions is $0 \cdot 212$. The positive and negative sampling errors seem more or less to cancel out, so that

$$E(p) = \Pi \qquad (11.2)$$

where $E(p)$ is the expected value of p. It follows that the expected value of the

(a)

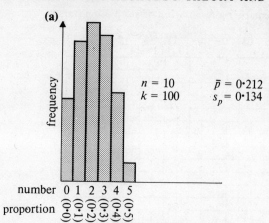

$n = 10$ $\bar{p} = 0.212$
$k = 100$ $s_p = 0.134$

Number of adults x_i	Frequency f_i
0	13
1	22
2	25
3	23
4	14
5	3
6	0
–	100

(b)

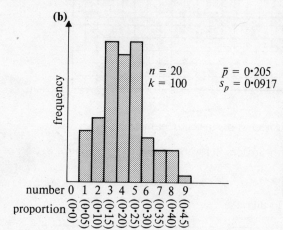

$n = 20$ $\bar{p} = 0.205$
$k = 100$ $s_p = 0.0917$

Number of adults x_i	Frequency f_i
0	0
1	8
2	10
3	22
4	20
5	22
6	7
7	5
8	5
9	1
10	0
–	100

(c)

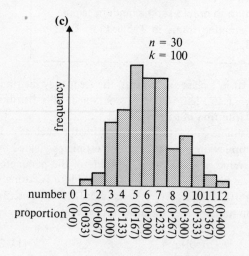

$n = 30$
$k = 100$

Number of adults x_i	Frequency f_i
0	0
1	1
2	2
3	10
4	12
5	19
6	17
7	17
8	6
9	8
10	5
11	2
12	1
–	100

Figure 11.1   Sampling distributions of p.

sampling error of p is

$$E(e_p) = E(p - \Pi) = 0 \tag{11.3}$$

Remember that it is only the expected value or long-run average that equals Π. The value of p obtained from any particular random sample may diverge, sometimes quite considerably, from Π (Fig. 11.1a). These formulae also show that the sampling distribution of e_p is identical in form to that of p. It is only the mean values which differ by an amount equal to the parameter being estimated.

> Assuming p is an unbiased estimator of $\Pi = 0\cdot2$, why should the mean of 100 estimates equal $0\cdot212$ rather than $0\cdot2$?

The probable magnitude of sampling error can be ascertained only if the standard deviation of p (and e_p) is known. Under certain conditions, the sampling distribution of p may be approximated by a normal distribution with mean or expected value given by

$$E(p) = \Pi \tag{11.4}$$

with a variance given by

$$\sigma_p^2 = \Pi(1 - \Pi)/n \tag{11.5}$$

and standard deviation or **standard error** given by

$$\sigma_p = \sqrt{\Pi(1 - \Pi)/n} \tag{11.6}$$

(It is customary to call the standard deviation of a sampling distribution a standard error.)

Let us examine the conditions under which the normal approximation holds by considering some other experimentally derived sampling distributions. We have seen that samples of size $n = 10$ give a skewed sampling distribution. Although there is still some evidence of skewness for $\Pi = 0\cdot2$ and $n = 20$ (Fig. 11.1b) a rather better approximation to normality is found. We may estimate the standard error of the sampling distribution for $\Pi = 0\cdot2$, $n = 20$ using grouped frequency data and the computational formula

$$s_p = \sqrt{[\Sigma f_i p_i^2 - (\Sigma f_i p_i)^2/k]/(k - 1)}$$

where the p_i are the experimentally derived sample proportions, and $k = 100$ is the number of samples (*not* to be confused with sample size, n). From the table accompanying Figure 11.1b

$$s_p = \sqrt{[5\cdot035 - (20\cdot5)^2/100]/99} = \sqrt{0\cdot832/99} = 0\cdot0917$$

By Equation 11.6, the theoretically derived value is

$$\sigma_p = \sqrt{\Pi(1-\Pi)/n} = \sqrt{0\cdot16/20} = 0\cdot0894$$

Actual and theoretical values are relatively close, differing only because of random sampling fluctuations. The approximation improves as sample size increases – the experimentally derived sampling distribution for $\Pi = 0\cdot2$, $n = 30$ and $k = 100$ is shown in Figure 11.1c.

> Use the frequency distribution accompanying Figure 11.1c to show that $s_p = 0\cdot074$. How does it compare with the value of the parameter σ_p? What is the estimate of the expected value of the p_i?

As a rule, the sampling distribution of p may be assumed to be approximately normal if both np and nq are greater than 5. A minimum sample size of 25 is generally advisable, no matter what the value of p. We should also add that the variance of the sampling distribution of p (and e_p) is minimised if estimates are based on simple random sampling, so that the estimate is the most precise we can hope for, given our sample size.

The results obtained above hold only if the assumptions of *simple random sampling* (or systematic sampling under conditions as described in Ch. 10) and *adequate sample size* are fulfilled. If we are trying to estimate the proportion of an area covered by a particular land use or soil type, by generating a simple spatial random or grid sample, we should remember that the spatial pattern of the phenomenon should show *no spatial autocorrelation*, otherwise standard errors will be overestimated (Switzer 1976).

Provided these assumptions are satisfied, the behaviour of a sample proportion, p, and of its associated sampling error, $e_p = (p - \Pi)$, may be predicted on a probabilistic basis because it is *normally distributed*. Thus, for $\Pi = 0\cdot2$, $n = 30$ we have

$$E(p) = \Pi = 0\cdot2 \qquad E(e_p) = E(p - \Pi) = 0\cdot0$$
$$\sigma_p = \sigma_{e_p} = \sqrt{0\cdot2 \times 0\cdot8/30} = 0\cdot073$$

where σ_{e_p} is the standard error of the sampling error, e_p, and, from Table A.3, $68\cdot26\%$ of the sample proportions should fall between $(\Pi \pm 1\sigma_p)$, i.e.

$$\text{prob}(\Pi - \sigma_p \leqslant p \leqslant \Pi + \sigma_p) = \text{prob}(0\cdot127 \leqslant p \leqslant 0\cdot273) = 0\cdot6826$$

or

$$\text{prob}(-\sigma_p \leqslant e_p \leqslant \sigma_p) = \text{prob}(-0\cdot073 \leqslant p \leqslant 0\cdot073) = 0\cdot6826$$

> Write down the limits within which p should fall (i) **95%** of the
> time, (ii) **99%** of the time. Do the same for e_p.

THE CENTRAL LIMIT THEOREM

The results demonstrated for sample proportions may be generalised to means
based on interval or ratio data by appealing to the Central Limit Theorem. This
theorem states that, no matter what the form of the frequency distribution of
the population of measurements for some variable X, the sampling distribution
of the sample mean $\bar{X}$ approaches a normal distribution as sample size n
increases. The more skewed the population of observations, the
larger the sample size has to be to ensure approximate normality.
If the population has mean μ and standard deviation σ, the mean or

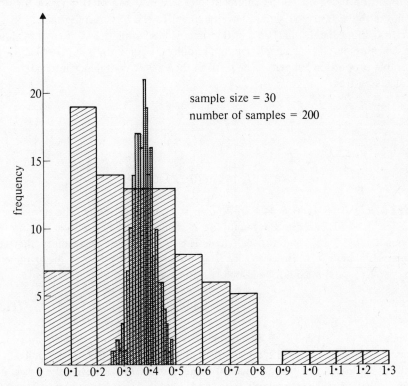

Figure 11.2   Relationship between population distribution and frequency distribution of sample mean values.

expected value of the sampling distribution of $\overline{X}$ will be $E(\overline{X}) = \mu$, and its standard error will be $\sigma_{\overline{X}} = \sigma/\sqrt{n}$. It is assumed that the estimate of $\overline{X}$ is based on measurements obtained in simple random sampling.

In a class experiment, 200 simple random samples of size 30 were drawn from a population of highway casualty rates per mile of road for 91 administrative regions in Great Britain. As Figure 11.2 shows, the population of measurements is strongly right-skewed. However, the frequency distribution of the 200 sample means, superimposed in the diagram, shows a close approximation to a normal distribution.

ESTIMATION AND CONFIDENCE INTERVALS

Confidence interval for Π

The theoretical results stated earlier are of great practical importance. When sampling, we do not know the value of Π (otherwise there would be no point sampling in the first place!) but only that of p. Taking the population of measurements listed in Table 11.1 for which $\Pi = 0.2$, we know that a sample proportion based on 30 measurements lies within 0.143 of Π, 95% of the time (this is the answer you should have obtained to the box exercise on p. 153). It follows immediately that, 95% of the time, Π lies within 0.143 of any individual proportion obtained in simple random sampling, assuming $n = 30$, $\Pi = 0.2$. If the sample proportion happens to be $p = 0.3$, then a 95% **confidence interval** may be calculated from

$$\Pi_{0.95} = p \pm 0.143$$

giving the interval

$$0.3 \pm 0.143 \text{ or } (0.157, 0.443)$$

The values 0.157 and 0.443 are known as **confidence limits**. The term 'confidence' is used because we cannot be *certain* that Π lies within the interval bounded by 0.157 and 0.443. There is a 95% or 0.95 probability that the interval contains Π. However, as Π is unknown the value of the parameter $\sigma_p = \sqrt{\Pi(1 - \Pi)/n}$ cannot be calculated so it is estimated by

$$s_p = \sqrt{pq/n} \tag{11.7}$$

where $q = (1 - p)$ and s_p is the *estimated* standard error of p. This may appear suspect, because if $\Pi = 0.2$, but $p = 0.3$, then $s_p = \sqrt{0.3 \times 0.7/30} = 0.084$ which differs (although not too drastically) from $\sigma_p = 0.073$. However, the procedure works well for adequate sample sizes and is unbiased in the long run, because the mean or expected value of s_p is σ_p, i.e. $E(s_p) = \sigma_p$. It is partly to reduce the

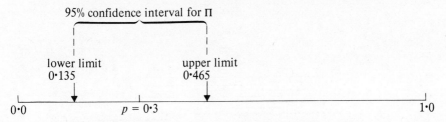

Figure 11.3   95% confidence interval for Π.

chance of serious error in estimating σ_p, as well as ensuring approximate normality, that a minimum sample size of 25 or 30 is recommended. Assuming $p = 0\cdot3$, the (estimated) 95% confidence interval may be written

$$\Pi_{0\cdot95} = p \pm 1\cdot96s_p = 0\cdot3 \pm (1\cdot96 \times 0\cdot084) = (0\cdot135, 0\cdot465)$$

The estimate p and its associated confidence interval are shown in Figure 11.3. Interpretation of this interval is as follows. The interval $(0\cdot135, 0\cdot465)$ represents only one of all the possible 95% confidence intervals that could have been constructed using a simple random sampling procedure based on a sample size of 30. Of these intervals, approximately 95% will contain the value of Π. It is very important to realise that the quantity being estimated is *fixed*, i.e. Π is 'deterministic', and the confidence interval is a 'random interval'. Different samples of the same size, drawn randomly from the same population, will yield varying values of p and, consequently, confidence intervals of varying location and slightly varying width. Figure 11.4 shows a set of 95% confidence intervals based on 100 simple random samples ($n = 30$) taken from the sampling frame in Table 11.1. Of these intervals four (or 4%) include $\Pi = 0\cdot2$, much as one would expect if sampling theory is correct. Notice also that 15 of the intervals include *negative* values, which is plainly nonsensical. Even with a sample size of 30, it is possible to obtain estimates of $\Pi = 0\cdot2$ such as $p = 1/30 = 0\cdot033$ or $p = 2/30 = 0\cdot067$ whose associated estimates of σ_p are relatively large. Hence the need to check that np and nq both exceed 5.

We are in a classic position of statistical inference, not knowing for certain whether any given confidence interval contains the value of the parameter or not. Furthermore, if the particular interval we have constructed does contain Π, there is no way of knowing exactly where Π falls within it. A 'good' confidence interval should be *precise*, or narrow relative to the magnitude of the quantity being estimated, and *reliable*, having a high probability of containing the unknown parameter value. These requirements may conflict. A 95% confidence interval is narrower, but less reliable, than the 99% confidence interval constructed from the same set of measurements. A 99% confidence interval is 'almost certain' to contain the parameter, but the width of the interval is greater than that at the 95% level.

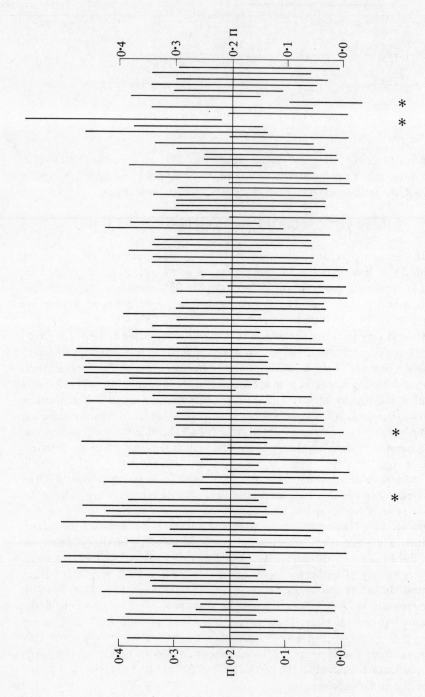

Figure 11.4   95% confidence intervals for $\Pi = 0.2$.

Notation

The general expression for a confidence interval is written in terms of α, where α represents the proportion of the area under two tails of the normal curve for a significance test carried out at the α or $\alpha \times 100\%$ significance level. The 95% confidence interval excludes the 5% most extreme sample proportions occurring in both tails of the normal distribution, so the confidence interval may be described as the $(1 - 0{\cdot}05)$ or $(1 - 0{\cdot}05) \times 100\%$ confidence interval, where $\alpha = 0{\cdot}05$. The critical value of the standard normal deviate z is that which defines an area under *one* tail of the normal curve of $\alpha/2$ or $(\alpha/2) \times 100\%$, and thus an area under *two* tails of the curve of α or $\alpha \times 100\%$ (Fig. 11.5). The $(1 - \alpha)$ or $(1 - \alpha) \times 100\%$ confidence interval for Π is therefore expressed by

$$\Pi_{1-\alpha} = p \pm (z_{\alpha/2} \times s_p). \tag{11.8}$$

For a 95% confidence interval, we write

$$\Pi_{0{\cdot}95} = \Pi_{1-0{\cdot}05} = p \pm (z_{0{\cdot}025} \times s_p) = p \pm 1{\cdot}96 s_p.$$

> Write down a similar expression for a 99% confidence interval for Π.

As in the cases of significance testing previously discussed, the investigator is not bound to keep to the conventional levels of $0{\cdot}95$ and $0{\cdot}99$.

Matowanyika (1977) made use of this technique in his undergraduate dissertation in which he hypothesised that the spatial distribution of West Indian immigrants in Britain depends chiefly upon internal migration. If this hypothesis were correct in the case of Reading, England, then the majority of West Indian households in the town should have previously lived in another British town. Of

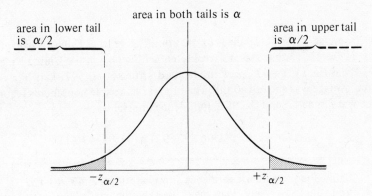

area in both tails is α

area in lower tail is $\alpha/2$

area in upper tail is $\alpha/2$

$-z_{\alpha/2}$ $+z_{\alpha/2}$

Figure 11.5   Region of rejection for significance test based on a sample proportion.

140 randomly selected households, 56 ($p = 0·4$) had lived in other towns, but 84 ($q = 0·6$) had not. These estimates suggest rejection of the hypothesis, but it is wise to check the probable size of the error involved. Since $np = 56$ and $nq = 84$ are both above 5, the 95% confidence interval may be calculated as

$$\Pi_{0·95} = 0·4 ± (1·96 \times 0·0414) = (0·319, 0·481)$$

This interval does not include the critical value of $0·5$, so we should be correct, 95% of the time, in rejecting the hypothesis of internal migration, at least in the Reading case.

> Calculate 99% confidence limits for Matowanyika's data. What would a 'reasonable person' conclude about the internal migration hypothesis on this evidence?

Confidence interval for μ

For *large samples*, in which $n \geqslant 30$, the sampling distribution of $\overline{X}$ is assumed to follow a normal distribution in accordance with the Central Limit Theorem. The general expression for obtaining the limits of a confidence interval is

$$\mu_{1-\alpha} = \overline{X} ± (z_{\alpha/2} \times s_{\overline{X}}) \tag{11.9}$$

Notice that the parameter $\sigma_{\overline{X}}$, the true standard error of $\overline{X}$, is estimated by $s_{\overline{X}}$ because the population standard deviation of X, namely σ, is usually unknown. Thus σ is estimated by

$$s = \sqrt{\frac{1}{n-1} \ [\Sigma X_i^2 - (\Sigma X_i)^2 / n]} \tag{11.10}$$

and

$$s_{\overline{X}} = s/\sqrt{n} \tag{11.11}$$

These formulae may be applied to the findings of the *Family Expenditure Survey 1974* (1975) that average weekly earned income in the United Kingdom in 1974 was (in £) $\overline{X} = 42·25$, with standard error $s_{\overline{X}} = 0·592$. Is it likely that the true average was above £40·00? The number of sample households recording income was $n = 4847$, and the 95% confidence interval is

$$\mu_{0·95} = 42·25 ± (1·96 \times 0·592) = (41·09, 43·41)$$

> Show that the 99% confidence limits are (40·72, 43·78). Do you think it likely that mean weekly income is less than £40·00?

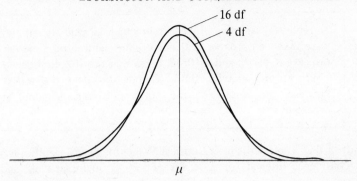

Figure 11.6 t distributions with different numbers of degrees of freedom.

For *small samples*, where $n < 30$, the sampling distribution of $\overline{X}$ is no longer normal, but follows a *t* distribution. This distribution is continuous, symmetrical about its mean, with an overall shape and standard error dependent upon the number of degrees of freedom, given by the sample size minus one, i.e. df = $(n - 1)$ (Fig. 11.6). However, $\overline{X}$ follows a t distribution only if the population from which the sample measurements were obtained is normally distributed. Sample observations should be checked to see whether they are likely to have been drawn from a normal population (see pp. 88–90), and/or a reasonable *a priori* case made for normality.

Confidence limits are obtained from

$$\mu_{1-\alpha} = \overline{X} \pm (t_{n-1,\alpha/2} \times s_{\overline{X}}) \qquad (11.12)$$

where $t_{n-1,\alpha/2}$ represents the number of standard error units we should move away from the mean of a distribution with $(n - 1)$ df such that $\alpha \times 100\%$ of the area falls in the two tails of the distribution and, hence, $(1 - \alpha) \times 100\%$ of the area lies symmetrically about the mean (Fig. 11.6). Values of t for various values of α and degrees of freedom are given in Table A.4.

Ball and Williams (1968) compared the errors in estimation of exchangeable cations of potassium (in mg/100 g) obtained from two soil-sampling procedures of an upland site on uncultivated, unfertilised brown earths in Snowdonia. A 'bulk-sampling' procedure, in which 22 sub-samples were taken from a single large sample block, gave a sample mean of $\overline{X} = 0\cdot55$ and sample standard deviation $s = 0\cdot026$. Thus $s_{\overline{X}} = 0\cdot026/\sqrt{22} = 0\cdot006$, and 95% limits for μ are

$$\mu_{0\cdot95} = \overline{X} \pm (t_{21,0\cdot025} \times 0\cdot006) = 0\cdot55 \pm (2\cdot08 \times 0\cdot006) = (0\cdot538, 0\cdot562)$$

So 95% of the time, an error within $\pm 0\cdot012$ is to be expected. The required value of t is found at the intersection of the df = $(n - 1) = 21$ row and $\alpha = 0\cdot025$ column of Table A.4.

A 'replicate' sampling procedure in which 22 separate samples were taken on a circle of 0·75 m radius centred on a marker peg gave $\bar{X} = 0·73$ and $s = 0·32$. Show that the 95% confidence limits for μ are (0·59, 0·87). Comment on the differences between the results of the bulk and replicate sampling procedures.

Notice that Table A.4 shows a critical t value for the 5% significance level (or 95% confidence interval) of 2·04 for df = 30, 2·00 for df = 60, and a value identical to that for a normal distribution (1·96) for an infinite (∞) number of degrees of freedom. Little error is involved in substituting a z value for a t value if n ⩾ 30.

SAMPLE SIZE AND SAMPLING ERROR

As Equations 11.8 and 11.9 show, the magnitude of sampling error is deter- mined both by the standard deviation of the population from which the sample measurements were drawn, and by sample size, n. As the investigator is far more likely to be able to influence the latter than the former, we shall concentrate on

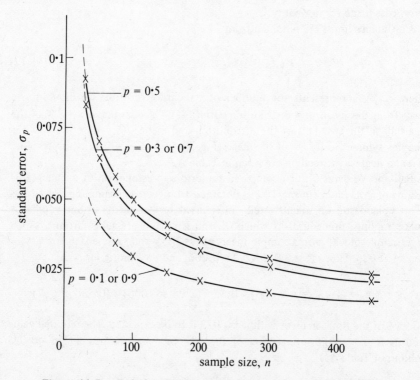

Figure 11.7   Relationship between standard error and sample size.

the relationship between sampling error, as expressed in terms of standard error, and sample size.

For both proportions and means, standard error is inversely proportional to $\sqrt{n}$ as illustrated for three sets of values of p in Figure 11.7. Notice that doubling or tripling sample size does not double or triple the precision of the estimate. Holding p constant, doubling n from 50 to 100, or trebling n from 50 to 150, respectively, produces standard errors approximately 70% or 58% of the original. To halve the standard error (or double precision), the original sample size must be multiplied by 2^2 to obtain $n = 200$; a tripling of precision is achieved by multiplying n by 3^2 to obtain 450. An absolute increase in sample size by a fixed amount is far more worthwhile if the initial sample size is quite small. An increase from 20 to 80 doubles precision but involves only 60 additional observations, whereas the same relative increase in precision, given an initial sample size of 100 requires an absolute increase of 300 ($=(100 \times 2^2) - 100$). The effect of varying p is to decrease the standard error as p diverges from $0·5$, since for $p = 0·5$, $pq = 0·25$; for $p = 0·3$ or $0·7$, $pq = 0·21$; and for $p = 0·9$ or $0·1$, $pq = 0·09$ (Fig. 11.7).

For sample means based on interval or ratio data, the relationship between standard error and n would look very similar. The **sampling fraction**, f, which is the ratio of sample size to population size, only has a noticeable effect on the standard error if it is 10% or more.

The sampling fraction should always be expressed as a decimal, i.e. a sampling fraction of 20% is $f = 0·2$, when used to correct the standard error formulae given previously. For a sample proportion, p, we have

$$s_p = \sqrt{(1-f)}\,\sqrt{pq/n} \qquad (11.13)$$

and for a sample mean, $\overline{X}$,

$$s_{\overline{X}} = \sqrt{(1-f)}\,(s/\sqrt{n}) \qquad (11.14)$$

where $(1-f)$ is called the **finite population correction**. The effect of the correction factor is to increase the precision of the estimate. If the sampling fraction in Matowanyika's study had been 10%, the standard error would have been

$$s_p = \sqrt{(1-0·1)}\,\sqrt{0·24/140} = 0·95 \times 0·0414 = 0·0393$$

compared with the uncorrected value of $0·0414$. Even for $f = 0·1$, the correction factor of $0·95$ makes little difference, explaining why it is usually ignored if $f < 0·1$. For $f = 1·0$, $s_p = 0$, which makes sense because it means the 'sample' consists of all measurements in the population!

> Establish 99% confidence limits for the proportion of West Indian households who previously lived in other British towns, assuming the sample size of $n = 140$ represents 20% of all West Indian households in Reading. Does the result alter your opinion of the internal migration hypothesis?

Although the correction factor may be used in the context of estimation in order to establish confidence limits, it *must not* be used to calculate standard errors for statistical tests of significance.

WORKING BACKWARDS – ESTIMATING REQUIRED SAMPLE SIZE

Some studies are of little value unless results attain a prespecified accuracy. For instance, it is difficult to gauge public reaction to a proposed shopping centre if estimates are subject to errors of 20–30%.

For measurements based on *nominal* or *count* data, the sample size required to yield the maximum specified margin of error e at the $(1 - \alpha)$ confidence level is given by rearranging the standard error component of Equation 11.8 to obtain

$$n = pq \times (z_{\alpha/2}/e)^2. \tag{11.15}$$

An estimate of p is required *before* n can be determined, and may be obtained in several ways.

(a) By referring to the results of previous similar studies, e.g. experience shows the proportion of voters for the Conservative Party at by-elections usually falls between 30 and 50%, so take $p = 0.4$.
(b) By carrying out a pilot survey based on a sample size of at least 30 and estimating p from this survey.
(c) By making an intelligent guess. The proportion of two-car households in a working-class area would almost certainly be less than 10%.
(d) By 'playing safe'. The expression (11.15) is maximised for $p = q = 0.5$, so choosing $p = 0.5$ means that the desired margin of error cannot be exceeded. However, this approach requires more time and effort.

If the proportion of shoppers in favour of a particular scheme is thought to be approximately $p = 0.7$ and an error of ± 0.10 (10%) can be tolerated at the 95% confidence level, then

$$n = 0.7 \times 0.3 \left(\frac{1.96}{0.1} \right)^2 = 0.21 \times 384 = 81$$

Provided the final sample size is to be considerably greater than 30 (to justify a pilot survey), the expression used to estimate sample size for *interval* or *ratio* measurements is

$$n = \left(\frac{z_{\alpha/2} \times s}{e} \right)^2 \tag{11.16}$$

where e is the maximum tolerable sampling error, s an estimate or guess of the standard deviation of the population of measurements, and $z_{\alpha/2}$ the critical value of a standard normal deviate corresponding to the $(1 - \alpha)$ confidence level. If a pilot survey is used, its sample size should be at least 30.

For many surveys in physical geography, a procedure involving small-sample pilot studies and based on the t distribution may be helpful. The critical sample size is given by

$$n = \left(\frac{t_{n-1,\,\alpha/2} \times s}{e} \right)^2 \qquad (11.17)$$

and is a formula to be handled with care as the value of t itself depends upon n. Ball and Williams (1968) determined sample size, giving sample mean values with errors of a specified magnitude at the 95% confidence level from a pilot survey whose sample size was $n = 22$ (Table 11.2). Let us determine n for estimating mean pH in a field-moist soil. From Table 11.2 we find $\bar{X} = 4\cdot9$, $s = 0\cdot17$. If we specify a maximum error of 2% relative to the sample mean, this becomes $e = \bar{X} \times 0\cdot02 = 0\cdot098$ in terms of the pH scale. To simplify the initial arithmetic, choose $t_{n-1,\alpha/2} = 2\cdot0$ and substitute in Formula 11.17 to obtain

$$n = \left(\frac{2\cdot0 \times 0\cdot17}{0\cdot098} \right)^2 = (3\cdot47)^2 = 12(\cdot04)$$

Taking our new value of $n = 12$ gives df $= (n - 1) = 11$ and $t_{11,0\cdot025} = 2\cdot20$ so that

$$n = \left(\frac{2\cdot2 \times 0\cdot17}{0\cdot098} \right)^2 = (3\cdot82)^2 = 14\cdot59 \simeq 15$$

Our answer now differs only slightly from that given by Ball and Williams, and taking df $= (15 - 1) = 14, t_{14,0\cdot025} = 2\cdot15$ gives

$$n = \left(\frac{2\cdot15 \times 0\cdot17}{0\cdot098} \right)^2 = (3\cdot73)^2 = 13\cdot91 \simeq 14$$

as shown in Table 11.2.

> Substitute the value for $t_{13,0.025}$ in Equation 11.17 and show that there is no justification for further calculations. Check some of the other sample sizes given in Table 11.2.

Table 11.2 ilustrates why it is rather meaningless to talk of the ideal sample size for any particular study, as those specified vary from 3 for the determination of exchangeable sodium (Na) cations to 102 for that of extractable phosphate (P_2O_5).

Table 11.2 Means and standard deviations of soil sample properties obtained from two areas in Snowdonia (adapted from Ball and Williams 1968).

Soil property	Location					
	Moel Eilio			Llydaw		
	Mean ($\bar{X}$)	Standard deviation (s)	Sample size (n)	Mean ($\bar{X}$)	Standard deviation (s)	Sample size (n)
pH						
Field-moist soil	4·9	0·17	14	5·0	0·12	5
Air-dry soil	4·5	0·12	9	4·7	0·16	12
Loss on ignition						
Weight per cent	19·5	1·65	13	12·2	1·56	30
Exchangeable cations (mg/100 g)						
K	0·73	0·320	70	0·34	0·120	57
Na	0·61	0·039	3	0·56	0·039	3
Ca	0·99	0·275	34	0·89	0·234	30
Mg	1·03	0·310	43	0·71	0·181	30
Mn	0·48	0·158	45	0·21	0·055	34
Extractable P_2O_5 (mg/100 g)	1·4	0·67	102	4·9	1·41	37

EXERCISES

1. Matthews (1975) wished to date four former ice margins of unknown age, using lichen measurements. A random sample of size $n = 24$ is taken from one of the margins, and lichen analysis gives a mean date of $\bar{X} = 1849$ and an estimated standard deviation $s = 19\cdot0$. What size of error would you expect, 95% of the time, in estimating the data of deglaciation at this margin? How large a sample is required to give an error margin of at most five years with 99% confidence?
2. The proportion of households in the suburbs of a large city who do their major weekly food shopping in the city centre is thought to be $p = 0\cdot30$. It is also thought that the standard deviation of the expenditure (in £) made on such trips is $3\cdot50$. How large a sample is required if the true proportion, II, and the mean expenditure, μ, are to be estimated with sampling errors of $\pm0\cdot05$ and $\pm1\cdot00$ respectively, at the 95% confidence level, on the same survey? Is there an 'ideal' sampling size? Show that error liable to occur at the 95% confidence level, estimating expenditure, is of the order of $0\cdot38$ if the larger sample size is employed.

REFERENCES AND FURTHER READING

A very clear discussion of the relationship between confidence intervals and significance tests is given in:
Bulmer, M. G. 1967. *Principles of statistics*, 2nd edn. Edinburgh: Oliver and Boyd. (pp. 165–9).
A thorough treatment of standard errors in a survey sampling context is given in:
Moser, C. A. and G. Kalton 1971. *Survey methods in social investigation*, 2nd edn. London: Heinemann.
Error estimates or confidence intervals are given in:
Lloyd, R. E. 1977. Consumer behaviour after migration: a reassessment process. *Econ. Geog.* 53(1), 15–27.
Simon, N. W. H. 1977. The relative level and changes in earnings in London and Great Britain. [sic]. *Reg. Stud.* 11(2), 87–98.
There are rather more examples in physical geography, as in:
Ball, D. F. and G. M. Williams 1968. Variability of soil chemical properties in two uncultivated brown earths. *J. Soil Sci.* 19(2), 379–91.
Carson, M. A. 1967. The magnitude of variability in samples of certain geomorphic characteristics drawn from valley-side slopes. *J. Geol.* 75, 93–100.
Matthews, J. A. 1975. Experiments on the reproducibility and reliability of lichenometric dates, Storbreen gletschervorfeld, Jotunheimen, Norway. *Norsk. Geog. Tidsskr.* 29(3), 97–109.
Strahler, A. N. 1950. Equilibrium theory of erosional slopes approached by frequency distribution analysis. *Am. J. Sci.* 248, 673–96, 800–14.
The data on migrant behaviour were taken from:
Matowanyika, J. Z. Z. 1977. *The migration behaviour of West Indian immigrants to Reading*. University of Reading, Dept of Geography, Unpublished undergraduate dissertation.
and those on family expenditure from:
Family expenditure survey, report for 1974 1975. London: H.M.S.O., Dept of Employment.

Comments on the effects of spatial autocorrelation on estimation of proportions are given in:

Switzer, P. 1976. Applications of random process models to the description of spatial distribution of qualitative geologic variables. In *Random processes in geology*, D. F. Merriam (ed.), 124—34. New York: Springer—Verlag.

12 Parametric Methods II: Hypothesis Testing

INTRODUCTION

Hypothesis testing in an artificial sampling context is used primarily to compare sample proportions or sample means calculated for two or more groups or areas, as illustrated by Questions 3, 9 and 10 in Table 10.1. Ideally, such comparisons will have been suggested by previous theorising, but in many cases striking differences may be discovered which then demand an explanation from the geographer. Whatever the reason for the comparison, if the data on which it is based are obtained by random sampling, then observed differences may be due to the vagaries of the sampling procedure rather than to any real difference between the populations of measurements. Under such circumstances, any conclusions should be stated in probabilistic terms based on the logic of statistical hypothesis testing. The tests described in this, and subsequent chapters, may also be applied in a natural sampling context if we view the sample means as produced by a stochastic 'natural' or 'socio-economic' process.

COMPARISON OF TWO SAMPLE MEANS

Proportions

For nominal or count data, the *proportions* of individuals or items in two different areas are compared. Matowanyika distinguished households migrating directly to Reading from the West Indies (group 1) from those migrating from within Great Britain (group 2), and argued that the proportions of households who stated that they knew someone in Reading before arrival would differ between the two groups. Assuming no prior expectations about the result, we carry out a *two-tailed* test based on the statistical hypotheses

$$H_0 : \Pi_1 = \Pi_2 \ (= \Pi)$$

$$H_1 : \Pi_1 \neq \Pi_2$$

These hypotheses must be stated correctly, otherwise errors will be made in the calculations. In effect, H_0 asserts that the difference $(\Pi_1 - \Pi_2)$ is equal to zero. If H_0 is true, the effect of repeatedly drawing pairs of samples of size n_1 and n_2, and calculating the difference between each pair, will be to obtain a sampling distribution of the difference $(p_1 - p_2)$. The expected value of this difference will be

$$E(p_1 - p_2) = 0$$

with the estimated variance of this difference given by

$$s_{p_1 - p_2}^2 = s_{p_1}^2 + s_{p_2}^2$$

where $s_{p_1}^2$ and $s_{p_2}^2$ estimate the variances of the sampling distributions of p_1 and p_2 respectively. The estimated standard error of the difference is calculated from

$$s_{p_1 - p_2} = \sqrt{s_{p_1}^2 + s_{p_2}^2} \qquad (12.1)$$

Under H_0, it is assumed that $\Pi_1 = \Pi_2 = \Pi$, so the true variances of the sampling distributions of p_1 and p_2 will each be given by $\sigma_p^2 = \Pi(1 - \Pi)/n$. The standard error of the difference must therefore be estimated in accordance with this assumption by *pooling* the estimate of σ_p^2 and, hence, σ_p. Letting $m_1 = 46$ be the number migrating directly with prior contacts, $n_1 = 56$ be the total number migrating directly, $m_2 = 79$ be the number migrating internally with prior contacts, and $n_2 = 84$ be the total number migrating internally, an estimate of Π, disregarding group membership, is given by

$$p = \frac{m_1 + m_2}{n_1 + n_2} = \frac{46 + 79}{56 + 84} = \frac{125}{140} = 0{\cdot}893$$

and the estimated standard error of $(p_1 - p_2)$ by

$$s_{p_1 - p_2} = \sqrt{s_p^2 + s_p^2} = \sqrt{pq \left(\frac{1}{n_1} + \frac{1}{n_2} \right)} \qquad (12.2)$$

The sample quantity $(p_1 - p_2)$ can be taken as approximately normally distributed, with standard error estimated by Equation 12.2, if pn_1, qn_1, pn_2, qn_2 are all greater than 5. This is the case here, so taking $p_1 = 46/56 = 0{\cdot}821$, $p_2 = 79/84 = 0{\cdot}940$ gives

$$z = \frac{|\, 0{\cdot}821 - 0{\cdot}940 \,|}{\sqrt{0{\cdot}893 \times 0{\cdot}107 \left(\frac{1}{56} + \frac{1}{84} \right)}} = \frac{0{\cdot}119}{0{\cdot}053} = 2{\cdot}25$$

(The absolute value (| |) is used because we ignore the sign for a two-tailed test.)

> What do you conclude from this result? Suppose you had been asked to predict the direction of the difference $(p_1 - p_2)$ before gathering the data. How would you restate the statistical hypotheses? What would you conclude?

Sometimes, we may wish to decide whether a sample proportion p has been obtained by drawing observations randomly from a population with a specified mean value Π. For example, Matowanyika's internal migration hypothesis could also be examined using a significance test based on the hypotheses

$$H_0: \; \Pi = 0\cdot5$$
$$H_1: \; \Pi > 0\cdot5$$

the true standard error of p being given by $\sqrt{\Pi(1 - \Pi)/n}$ (since H_0 assumes a particular value for Π) and the test statistic by

$$z = \frac{p - \Pi}{\sqrt{\Pi(1 - \Pi)/n}} \tag{12.3}$$

As $p = 0\cdot4$ in Matowanyika's study, falling on the 'wrong side' of $\Pi = 0\cdot5$, we cannot really come to a satisfactory conclusion in terms of the statistical hypotheses given above. An approach based on confidence limits as outlined in the previous chapter is often more helpful, showing how estimation is linked to, and aids, hypothesis testing.

Means

For measurements based on interval or ratio data, comparison of two (arithmetic) sample means, $\overline{X}_1$ and $\overline{X}_2$, is appropriate.

First, we consider *large samples*, for which the sample sizes in each area or group should be 30 or more. In a study of student knowledge of Reading town centre (Ratcliff *et al* 1977) it was hypothesised that second-year students would show greater knowledge than first-year students, as measured by their ability to locate 36 photographs of town centre street scenes correctly on a map. The statistical hypotheses are

$$H_0: \; \mu_1 = \mu_2$$
$$H_1: \; \mu_1 < \mu_2$$

where μ_2 and μ_1 respectively represent mean knowledge scores of all second- and all first-year undergraduates at Reading University. A one-tailed test is appropriate here. Responses were obtained from $n_1 = 30$ and $n_2 = 29$ randomly selected students, the sample means being $\overline{X}_1 = 14\cdot87$, $\overline{X}_2 = 19\cdot98$, and the sample variances $s_1^2 = 56\cdot16$ $s_2^2 = 65\cdot05$. The test statistic is

$$z = \frac{\overline{X}_1 - \overline{X}_2}{\sqrt{(s_1^2/n_1) + (s_2^2/n_2)}} \tag{12.4}$$

and the expression in the denominator is $s_{\overline{X}_1 - \overline{X}_2}$ or the estimated standard error of (the sampling distribution of) the difference between two sample means based on n_1 and n_2 observations, respectively. Substituting in Equation 12.4 gives

$$z = \frac{14\cdot87 - 19\cdot98}{\sqrt{56\cdot16/30 + 65\cdot05/29}} = \frac{-5\cdot11}{\sqrt{1\cdot872 + 2\cdot243}} = -2.52$$

> Use Table A.3 to confirm that this result is significant at the
> 1% level. Would it be significant at this level if the test were
> two-tailed?

Although, strictly speaking, both n_1 and n_2 should be 30 or more, the value
of 29 for n_2 in this case is close enough to be acceptable.

Occasionally, comparison of an individual sample mean $\bar{X}$ with a hypothesised population mean μ^* may be required. We set up the statistical hypotheses

$$H_0: \ \mu = \mu^*$$
$$H_1: \ \mu \neq \mu^*$$

for a two-tailed test, and modify H_1 for a one-tailed test. The test statistic is

$$z = \frac{\bar{X} - \mu^*}{s/\sqrt{n}} \tag{12.5}$$

where $s_{\bar{X}} = s/\sqrt{n}$ is the estimated standard error of $\bar{X}$.

To compare the means of *small samples*, the t test, based on the t distribution, is used. It is assumed that:

(a)   the populations from which the sample observations are obtained are
        normally distributed, and
(b)   the variances of the two populations are equal — this is also known as the
        homoscedasticity assumption.

Checks on these assumptions should be made before the test is carried out, but
these are deferred in the interests of clarity.

Bowlby (1972) used a psychological scaling technique to ascertain how
familiar the residents of a small city block in Evanston, Illinois, felt themselves
to be with each of a randomly selected sample of stores. It was thought that
perceived familiarity might vary with direction, and the scores for $n_1 = 16$ stores
lying north and $n_2 = 11$ stores lying south of the city block are given in Table
12.1. The scores are measured on an interval scale, and the higher the score the
greater the perceived familiarity. Assuming no prior expectations about the
direction of the difference between the mean familiarity scores for the two
sectors, the hypotheses are

$$H_0: \ \mu_1 = \mu_2$$
$$H_1: \ \mu_1 \neq \mu_2$$

where μ_1 and μ_2 are the means for stores in the north and south sectors,
respectively. In accordance with the equal variance assumption, a *pooled* estimate of the variance based on both samples is calculated from

$$s^2 = \frac{1}{n_1 + n_2 - 2} \{ [\Sigma X_{1j}^2 - (\Sigma X_{1j})^2/n_1] + [\Sigma X_{2j}^2 - (\Sigma X_{2j})^2/n_2] \} \tag{12.6}$$

Table 12.1   Observations for store familiarity study.

'Northern' stores		'Southern' stores	
X_{1i}	X_{1i}^2	X_{2i}	X_{2i}^2
13·0	169·0	6·4	41·0
10·6	112·4	10·2	104·0
9·2	82·8	9·1	82·8
11·4	130·0	8·3	68·9
8·5	72·3	6·7	44·9
10·3	106·1	7·3	53·3
11·9	141·6	8·1	65·6
6·1	37·2	1·7	2·9
8·8	77·4	5·8	33·6
10·2	104·0	5·1	26·0
10·1	102·0	6·9	47·6
10·0	100·0		
8·0	64·0		
4·9	24·0		
8·8	77·0		
9·4	88·0		
151·2	1487·8	75·6	570·6

$$n_1 = 16 \qquad\qquad n_2 = 11$$
$$\bar{X}_1 = 9\cdot45 \qquad\qquad \bar{X}_2 = 6\cdot87.$$

The estimated standard error of the difference is given by

$$s_{\bar{X}_1 - \bar{X}_2} = s\sqrt{\frac{1}{n_1} + \frac{1}{n_2}}. \qquad (12.7)$$

Substituting, we obtain

$$s_{\bar{X}_1 - \bar{X}_2} = \sqrt{4\cdot399}\sqrt{\frac{1}{16} + \frac{1}{11}} = 2\cdot097 \times 0\cdot392 = 0\cdot822$$

The test statistic is

$$t_{n_1 + n_2 - 2} = \frac{\bar{X}_1 - \bar{X}_2}{s\sqrt{\frac{1}{n_1} + \frac{1}{n_2}}} \qquad (12.8)$$

or

$$t_{25} = \frac{9\cdot45 - 6\cdot87}{0\cdot822} = 3\cdot14$$

From Table A.4 the critical 5% and 1% points for a t distribution with 25 df are 2·06 and 2·79 so that we may reject H_0 at the 1% significance level. Bowlby accounted for the difference by suggesting that the largely middle-class respondents were more likely to shop in the middle-class suburbs north of Evanston rather than at stores on and just south of the Chicago city boundary.

Checking assumptions

Checks for normality were described in Chapter 7. The equal variance (homoscedasticity) assumption may be checked by carrying out an F test upon the ratio of the sample variances obtained from the two sets of observations. If the two population variances are equal, it can be shown that the ratio of the two sample variances, based upon sample sizes of n_1 and n_2, follow the F distribution with $\nu_1 = (n_1 - 1)$ and $\nu_2 = (n_2 - 1)$ degrees of freedom (Fig. 12.1). To carry out the test, set up the hypotheses

$$H_0 : \sigma_1^2 = \sigma_2^2 \ (= \sigma^2)$$

$$H_1 : \sigma_1^2 \neq \sigma_2^2$$

A *two-tailed* test is required because we normally have no prior knowledge of the relative magnitude of the population variances. Because the critical values of F provided in Table A.5 are for a one-tailed test, the critical values for a *two-tailed* test at the α significance level are found by consulting the table of critical values at the $\alpha/2$ level for a one-tailed test. For example, the value for an F distribution with $\nu_1 = 10$ df and $\nu_2 = 15$ df at the 0·05 or 5% level is found, by interpolation, to be 3·07, since in Table A.5, $F_{10,14,0·025} = 3·15$ and $F_{10,16,0·025} = 2·99$. As this is the *upper tail* critical value, and the F distribution is not symmetrical (Fig. 12.1), the test statistic must be formed by dividing the greater variance estimate, s_G^2 by the smaller variance estimate, s_S^2. Using Formula 11.10 and the data in Table 12.1

$$s_S^2 = \tfrac{1}{15}[1487·8 - (151·2)^2/16] = 3·93$$
$$s_G^2 = \tfrac{1}{10}[570·6 - (75·6)^2/11] = 5·10$$

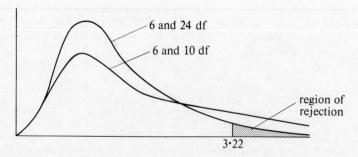

6 and 24 df

6 and 10 df

region of rejection

3·22

Figure 12.1   Examples of F distributions.

The general form of the test statistic is

$$F_{\nu_G, \nu_S} = s_G^2 / s_S^2 \qquad (12.9)$$

and for the familiarity data

$$F_{10,15} = \frac{5 \cdot 10}{3 \cdot 93} = 1 \cdot 30$$

Because this value falls below 3·07, and therefore not in the region of rejection of the kind shown in Figure 12.1, we decide to accept H_0 for lack of contrary evidence. In this case we were justified in pooling the variances and carrying out the t test. The F test also depends upon the assumption that measurements in the two populations are normally distributed.

For the t test, the crucial assumption is that of normality. However, it may be possible to fulfill this assumption by *transforming* 'badly behaved' observations so that they approximately follow a normal distribution (see pp. 179–80). If the number of observations in each sample is equal, the equal variance assumption is far less critical.

To compare an individual sample mean based on $n < 30$ observations with a hypothesised mean μ^*, the statistical hypotheses are as given for the large sample case on p. 170 and the test statistic is

$$t_{n-1} = \frac{\bar{X} - \mu^*}{s/\sqrt{n}} \qquad (12.10)$$

The assumption of normality must be fulfilled, but that of equal variances is no longer relevant.

We should also mention that if the sample observations used in the t test are spatially located and normally distributed, and also are spatially autocorrelated, then the estimated standard error for the test statistic should be corrected using the formula described by Cliff and Ord (1975) and discussed in Haggett *et al* (1977, pp. 330–4, 374–7).

COMPARISON OF THREE OR MORE SAMPLE MEANS – THE ANALYSIS OF VARIANCE (ANOVA)

Studies which depend upon statistical comparison of means would be unduly constrained if it were only possible to compare means two at a time. The **Analysis of Variance** allows us simultaneously to compare three or more sample means based on interval or ratio data, and the chi-square test provides the same facility with respect to three or more sample proportions based on nominal data as shown in Chapter 13.

Assumptions and theory

To make the necessary comparison, the problem is redefined so that an F test based on comparison of two sample variances may be used. The assumptions of random sampling, normality and equal variance underlying the technique are the same as for the t test.

If sample observations are obtained for k groups ($k \geqslant 3$), the statistical hypotheses are

$$H_0: \mu_1 = \mu_2 = \mu_3 = \cdots = \mu_k \ (=\mu)$$

$$H_1: \text{at least two of the } \mu_i \text{ are unequal}$$

(12.11)

Under H_0, for which the distribution of the (unknown) population of measurements can be as shown in Figure 12.2a, the frequency distributions of sample observations indicated schematically in Figure 12.2b are much more likely to occur in chance sampling than those in Figure 12.2c, in which differences between sample means are greater. The logic of the Analysis of Variance depends upon a comparison of the variation *between* groups, as measured by differences between the sample means, and the variation *within* groups, as measured by the amount of variation about each sample mean averaged over all groups.

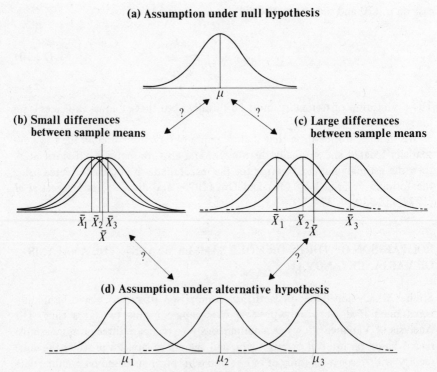

(a) Assumption under null hypothesis

(b) Small differences between sample means

(c) Large differences between sample means

(d) Assumption under alternative hypothesis

Figure 12.2   Hypothesis testing in the analysis of variance.

> Would the ratio of between-group variation to within-group
> variation be greater in Figure 12.2b or Figure 12.2c? Why?

The total variation in the sample observations may be expressed in terms of the **total sum of squares** (TSS), a quantity which in turn may be split into two components, the **between-group sum of squares** (BSS) and the **within-group sum of squares** (WSS), according to

$$\underset{i\ j}{\Sigma\Sigma}(X_{ij} - \bar{X})^2 = \underset{i}{\Sigma}n_i(\bar{X}_i - \bar{X})^2 + \underset{i\ j}{\Sigma\Sigma}(X_{ij} - \bar{X}_i)^2 \qquad (12.12)$$

$$\text{(TSS)} \qquad\qquad \text{(BSS)} \qquad\qquad \text{(WSS)}$$

Notice that TSS represents the sum of the squared deviations between each sample observation, X_{ij}, and the 'overall' or 'grand' sample mean of all the observations, $\bar{X}$.

> Use Figure 12.2 to show what $(\bar{X}_1 - \bar{X})$ and $(X_{1j} - \bar{X}_1)$
> represent. Then, state in words what BSS and WSS represent.

Still assuming H_0 to be true, the sample means $\bar{X}_1$, $\bar{X}_2$ and $\bar{X}_3$ all represent independent estimates of μ, and BSS provides an unbiased estimate of the variance, σ^2, of the population of observations shown in Figure 12.2a, in terms of the **between-group variance**

$$s_B^2 = \text{BSS}/(k - 1) \qquad (12.13)$$

where k is the number of groups, and $(k - 1)$ the degrees of freedom associated with this variance estimate. The quantity WSS provides another unbiased estimate (independent of that given by s_B^2) of σ^2 in terms of the **within-group variance**

$$s_W^2 = \text{WSS}/(n - k) \qquad (12.14)$$

where n is the total number of sample observations, and $(n - k)$ the degrees of freedom.

It can be shown that the ratio

$$F_{\nu_1, \nu_2} = s_B^2/s_W^2 \qquad (12.15)$$

follows an F distribution with ν_1 and ν_2 degrees of freedom when H_0 is true. For a problem in which there are ten observations in each group, and three groups, the degrees of freedom would be $\nu_1 = (3 - 1) = 2$ and $\nu_2 = (30 - 3) = 27$, and an F ratio exceeding $3 \cdot 35$ is to be expected at most 5% of the time. A calculated F ratio exceeding $5 \cdot 50$, which might be obtained with respect to the

Original observations

Class mid-values (sq. ft)	100	200	300	400	500	600	700	800	900	1000	1100
Frequency	6	14	7	6	5	3	2	0	1	0	1

Logarithmically transformed observations

Class mid-values (log sq.ft)	1·9	2·1	2·3	2·5	2·7	2·9	3·1
Frequency	2	7	11	12	7	5	1

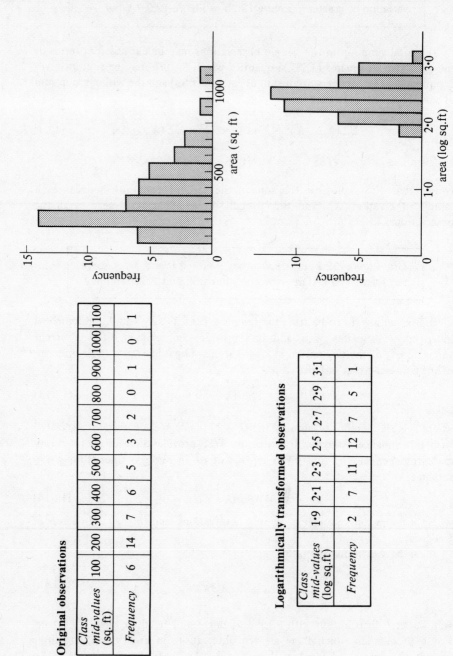

Figure 12.3   Original and transformed observations on second-order drainage basins.

EXAMPLE   177

three groups of sample observations shown in Figure 12.2c, would be expected at most 1% of the time (one-tailed test: see below) if H_0 were true, and so we might well prefer to believe that such observations were obtained from the populations of measurements shown in Figure 12.2d.

Intuitively it seems reasonable that the estimate of σ^2 given by the between-group variance s_B^2 is biased upward if H_0 is false. The estimate of σ^2 given by the within-group variance s_W^2 is unbiased no matter whether H_0 is true or false. To reject H_0, therefore, we always expect s_B^2 to exceed s_W^2 and so the F test must be *one-tailed* because we are interested only in the large ratios to be found in the right-hand tail of the distribution (Fig. 12.1).

By converting a 'comparison of means' problem into a 'comparison of variances' problem, the statistical hypotheses can now be stated

$$H_0 : E(s_B^2) = E(s_W^2) \, (= \sigma^2)$$

$$H_1 : E(s_B^2) > E(s_W^2) \tag{12.16}$$

where $E(s_W^2)$ and $E(s_B^2)$ are the expected values of the within- and between-group sample variances respectively, and σ^2 the true variance of the population(s) of measurements. The hypotheses stated in (12.16) are equivalent to those previously stated in (12.11).

Example

An investigation was made of the median particle sizes in soil parent materials near the Goring Gap (Thames Valley, England), which were thought to be derived from different sources. *Before* any mechanical or heavy mineral analysis was carried out, each of the 30 sample observations was placed either in group 1, thought to be derived from *in situ* weathering on the chalk, group 2, thought to be derived from aeolian additions to existing soils, or group 3, thought to be derived from fluvial deposits possibly originating as fluvio-glacial outwash in the the upper Thames valley (Table 12.2). All measurements are in μm (microns).

State the statistical hypotheses for this problem.

Computational formulae for TSS and BSS will be employed, and Equation 12.12 gives us an easy way to calculate WSS by subtraction, i.e.

$$WSS = TSS - BSS \tag{12.17}$$

Results of the calculations will be entered in an ANOVA table (Table 12.3). From Table 12.2, we find $\overline{X}_1 = 19 \cdot 6$, $\overline{X}_2 = 25 \cdot 0$, $\overline{X}_3 = 27 \cdot 8$ (all measured in microns) and $\overline{X} = 24 \cdot 1$. By the computational formulae, we have

$$TSS = \sum_i \sum_j X_{ij}^2 - n\overline{X}^2 \tag{12.18}$$

Table 12.2   Observations on median particle sizes in drift deposits.

Source 1		Source 2		Source 3	
X_{1i}	X_{1i}^2	X_{2i}	X_{2i}^2	X_{3i}	X_{3i}^2
15	225	19	361	25	625
17	289	23	529	22	484
8	64	27	729	31	961
27	729	29	841	29	841
31	961	24	576	30	900
9	81	28	784	33	1089
16	256	27	729	33	1089
26	676	26	676	25	625
18	324	27	729	27	729
29	841	20	400	23	529
196	4446	250	6354	278	7872

$$n_1 = 10 \qquad\qquad n_2 = 10 \qquad\qquad n_3 = 10$$
$$\bar{X}_1 = 19\cdot 6 \qquad\quad \bar{X}_2 = 25\cdot 0 \qquad\quad \bar{X}_3 = 27\cdot 8.$$

$n = 30$, $\bar{X} = 24\cdot 1$, and $\Sigma\Sigma X_{ij}^2 = 4446 + 6354 + 7872 = 18672$.

$$BSS = \sum_i n_i \bar{X}_i^2 - n\bar{X}^2 \tag{12.19}$$

and for our example this gives

$TSS = 18672 - 30 \times (24\cdot 1)^2 = 1247\cdot 7$

$BSS = [10 \times (19\cdot 6)^2 + 10 \times (25\cdot 0)^2 + 10 \times (27\cdot 8)^2] - 17424\cdot 3 = 395\cdot 7$

$WSS = 1247\cdot 7 - 395\cdot 7 = 852\cdot 0$

These values are entered in the first column of Table 12.3. Degrees of freedom are entered in the second column of this table. For BSS, df $= \nu_1 = (k-1) = (3-1) = 2$, and for WSS, df $= \nu_2 = (n-k) = (30-3) = 27$.

Show that the calculated F ratio is 6·27, and confirm using Table A.5 that a significant difference exists between the means at the 1% level.

Table 12.3   Analysis of Variance for particle size data.

Source	Sum of squares (SS)	Degrees of freedom (df)	Variance or mean square (MS)	F ratio
Between	395·7	2	$s_B^2 = 197\cdot 85$	$F = s_B^2/s_W^2 = 6\cdot 27$
Within	852·0	27	$s_W^2 = 31\cdot 56$	
Total	1247·7	29		

From this evidence, it seems reasonable to conclude that the means of the median particle sizes are different. Such evidence alone is not sufficient to show that the deposits originated in different areas, but could be used in conjunction with that provided by heavy mineral analysis or till fabric analysis.

For a comparison of two means, the t and F tests are equivalent. In general, $F_{1,v_2,\alpha} = (t_{v_2,\alpha/2})^2$, and reference to Tables A.4 and A.5 will confirm this.

Because of the form of the alternative hypothesis H_1, the F test cannot tell us *which* pairs of population means are likely to be unequal. Pairs of sample means may be compared using the t test and the test statistic

$$t_{n_i + n_l - k} = \frac{\overline{X}_i - \overline{X}_l}{s_W \sqrt{\dfrac{1}{n_i} + \dfrac{1}{n_l}}} \qquad (12.20)$$

where $\overline{X}_i$ and $\overline{X}_l$ are the sample means in question, n_i and n_l the number of observations in groups i and l respectively, $n_i + n_l - k$ the degrees of freedom, and $s_W = \sqrt{s_W^2}$. However, the test should be used only when the pairs of comparisons concerned have been specified *before* any statistical analysis of the data have been undertaken. If this stricture is ignored, and the investigator uses a poorly planned or exploratory study to hunt for differences too large to be attributed to sampling, the probability of committing a type 1 error will be seriously underestimated.

TRANSFORMATIONS

Although the requirements of normality and equal variance may not be met in sample data, it is often possible to transform these data to allow parametric tests to proceed.

We can deal with sample observations which appear to have been drawn from unimodal but non-normal distributions. Socio-economic data on income, or length of the journey to work, and geomorphic data on drainage basin areas or river flood heights tend to be positively skewed. Logarithmic transformations of the original values are often successful. The frequency distribution of areas (in square feet) of second-order drainage basins at Perth Amboy (Schumm 1956) and associated histogram are shown in the upper half, and the corresponding items based on logarithms of the basin areas in the lower half, of Figure 12.3. The distribution of basin areas is said to be **lognormal**. If any of the original observations includes a negative value or zero, an appropriate constant should be added to every observation. This is because logarithms of negative values are undefined, and the logarithm of zero is minus infinity! For cases of milder skewness a square-root transformation should suffice.

Sample observations from negatively-skewed populations are rarer, and occur only if many observations are at or near an upper limit. This may apply to percentage data, e.g. percentage of households with television sets or with indoor baths. If the original values are subtracted from 100, thus changing the term 'percentage with facility X' to 'percentage without facility X', a positively skewed distribution is obtained. A logarithmic transformation may then be performed.

Transformations to ensure equal variance are also available, and Haggett (1964) employs an arcsin transformation for this purpose with respect to data on the proportion of forested area in different regions in Brazil. For a discussion, see Miller and Kahn (1962, p. 175).

We should mention that many researchers do not regard moderate departures from the assumptions underlying parametric tests (particularly those underlying use of t and F tests) as particularly serious. This point of view is well stated in Boneau (1960) and Nunnally (1967). Parametric techniques are relatively robust, therefore, although not as robust as the non-parametric techniques discussed in Chapter 13.

EXERCISES

1. Ball and Williams (1968) found the mean pH values in field-moist soil samples obtained from Moel Eilio and Llydaw in Snowdonia to be $\bar{X}_1 = 4 \cdot 9$ and $\bar{X}_2 = 5 \cdot 0$ respectively. You are given that $n_1 = n_2 = 22$, $s_1^2 = 0 \cdot 029$ and $s_2^2 = 0 \cdot 014$.

   What test should you use to compare the two sample means? Carry out the test and state whether the result is significant.

2. It has been generally held by some historical geographers that coalmining families were particularly large in Victorian England. A set of 10% sample data from the 1871 census returns for the Bolton area showed mean numbers of children for mining families to be $\bar{X}_1 = 3 \cdot 46$ and for cotton workers' families to be $\bar{X}_2 = 3 \cdot 14$. You are given that $s_1^2 = 0 \cdot 020$, $s_2^2 = 0 \cdot 016$, $n_1 = 213$, $n_2 = 250$. Do you agree with the opinion expressed by the historical geographers? What does the result of your test show?

3. In addition to the study of perceived familiarity with sample stores in Evanston described earlier (pp. 170–1), Bowlby investigated perceived use. The study was also extended to cover stores lying in a sector west of the residents of the city block as well as those to the north and south. The eastern sector covered Lake Michigan. A priori theorising suggested that use scores would be higher in the sector oriented towards Evanston's CBD (Central Business District) than in the other two. The southern sector contains the CBD. The results were as given in the table.

North sector	West sector	South sector	Overall
$\bar{X}_1 = 15 \cdot 59$	$\bar{X}_2 = 12 \cdot 00$	$\bar{X}_3 = 8 \cdot 19$	$\bar{X} = 11 \cdot 40$
$n_1 = 16$	$n_2 = 11$	$n_3 = 23$	

ANOVA table

Source	SS	df	MS	F
Between	511·60	2	255·800 ⎫	
Within	1640·56	47	34·906 ⎭	7·33
Total	2152·16	49		

Sample variances of the use scores for the sectors were $s_1^2 = 15 \cdot 14$, $s_2^2 = 48 \cdot 25$ and $s_3^2 = 39 \cdot 44$.

(a) Given the evidence, do you think the assumptions underlying the use of the Analysis of Variance are approximately fulfilled? Use tests of significance where appropriate.

(b) Assuming the assumptions to be approximately fulfilled (even if they are not), state the null and alternative hypotheses and carry out the test.

(c) Are you justified in comparing any individual pairs of means? (Think very carefully about the relationship between *a priori* theorising and the results actually obtained.)

4. Melton (1960) examined the effects of 'erosional environment' on slope angles in the Ragged-Top Mountain area, Wyoming. Three valleys were carefully selected to control for other factors such as bedrock, vegetation, etc. The slope observations for each valley were classified either as NF, where there was no alluvial fan in the neighbourhood of the slope; FF, where there was an alluvial fan facing the slope (i.e. opposite a tributary entering on the other side); or AF, where the slope was situated above an alluvial fan (i.e. very close to a tributary entering on the same side). The observations were as follows.

NF (no fan)	FF (facing fan)	AF (above fan)
19·12	19·38	16·88
18·38	20·25	14·50
21·00	23·17	18·62
14·75	16·44	13·50
12·38	15·50	13·25
15·33	16·33	14·00

$\bar{X}_1 = 16 \cdot 83 \qquad \bar{X}_2 = 18 \cdot 51 \qquad \bar{X}_3 = 15 \cdot 12$
$\bar{X} = 16 \cdot 82.$

(i) Is there anything noteworthy about the differences between the means, and their relationship to the overall mean, $\bar{X}$?

(ii) One of Melton's conclusions is that frost action, past or present, is not necessarily primarily responsible for valley asymmetry. Bearing in mind that half the observations in each erosional environment are on north-facing slopes, and half on south-facing slopes, i.e. controlling for aspect, what kind of analysis should be carried out to confirm or reject his conclusion? Do you agree with him?

REFERENCES AND FURTHER READING

Clearly presented examples of *t* and *z* tests are found in:

Strahler, A. N. 1950. Equilibrium theory of erosional slopes approached by frequency distribution analysis. *Am. J. Sci.* **248**, 673–96, 800–14.

Schumm, S. A. 1956. Evolution of drainage systems and slopes in badlands at Perth Amboy, New Jersey. *Bull. Geol. Soc. Am.* **67**(5), 597–646.

A battery of *t* tests is applied to aspects of soil, vegetation and micro-relief features by:

Mitchell, C. W. and S. G. Willimott 1974. Dayas of the Moroccan Sahara and other arid regions. *Geog. J.* **140**(3), 441–53.

For examples related to human geography, see:

Bowlby, S. R. and J. A. Silk 1978. *Learning by doing: behavioural and social geography projects.* Reading; University of Reading, Dept of Geography, Geographical Papers.

Brail, R. K. and F. S. Chapin (Jr) 1973. Activity patterns of urban residents. *Environ. Behav.* **5**(2), 163–90.

Doling J. 1976. The family life cycle and housing choice, *Urban Stud.* **13**, 55–8.

The analysis of variance has also been widely employed by physical geographers in:

Bryan, R. B. 1974. A simulated rainfall test for the prediction of soil erodibility. *Z. Geomorph.* (Supplement) **21**, 138–50.

Carter, C. A. and R. J. Chorley 1961. Early slope development in an expanding stream system. *Geol Mag.* **98**, 117–30.

Doornkamp, J. C. and C. A. M. King 1971. *Numerical analysis in geomorphology.* London: Edward Arnold.

Melton, M. A. 1960. Intravalley variation in slope angles related to microclimate and erosional environment. *Bull. Geol. Soc. Am.* **71**, 133–44.

For examples in human geography, see:

Herzog, T. R., S. Kaplan and R. Kaplan 1974. The prediction of preference for familiar urban places. *Environ. Behav.* **8**(4), 627–45.

Lloyd, R. E. 1977. Consumer behaviour after migration: a reassessment process. *Econ. Geog.* **53**(1), 15–27.

Murdie, R. A. 1969. *Factorial ecology of metropolitan Toronto, 1951–1961.* Chicago: University of Chicago, Dept of Geography, Research Paper no. 116.

Corrections for spatial autocorrelation in the standard *t* test are given in:

Cliff, A. D. and J. K. Ord 1975. The comparison of means when samples consist of spatially autocorrelated observations. *Environ. Plann.* A **7**, 725–34.

Haggett, P., A. D. Cliff and A. E. Frey 1977. *Locational analysis in human geography,* 2nd edn. Vol. 2: *Locational methods.* London: Edward Arnold.

Transformations of variables are considered by:

Miller, R. L. and J. S. Kahn 1962. *Statistical analysis in geological sciences.* New York: Wiley.

Examples are given in Schumm (1956) (referenced above) and:

Haggett, P. 1964. Regional and local components in the distribution of forested areas in southeast Brazil: a multivariate approach. *Geog. J.* **130**, 365–80.

Violation of test assumptions is considered in:

Boneau, C. 1960. The effects of violations of assumptions underlying the *t* test. *Psychol. Bull.* **57**, 49–64.

Nunnally, J. 1967. *Psychometric theory.* New York: McGraw-Hill.

For an introductory discussion on the robustness of parametric techniques see:

Norcliffe, G. B. 1977. *Inferential statistics for geographers.* London: Hutchinson.

Other sources for examples and exercises were:

Ball, D. F. and G. M. Williams 1968. Variability of soil chemical properties in two uncultivated brown earths. *J. Soil Sci.* **75**, 93–100.

Bowlby, S. R. 1972. *Spatial variation in consumers' information levels.* Evanston: Northwestern University, The Transportation Centre, Household Travel Behaviour Study Report no. 4.

Ratcliff, H. M., A. P. Pilditch and N. R. Walter 1977. *A study of students' knowledge of Reading town centre.* University of Reading, Dept of Geography, Unpublished Behavioural Geography Project.

13   Non-Parametric Methods

INTRODUCTION

The number of assumptions that have to be fulfilled, and checked, particularly for procedures based on the t and F distributions, can restrict the use of parametric methods, and calculations may be rather involved and tedious, especially when transformations are introduced. Statistical procedures have been developed which are known as **non-parametric** or **distribution-free** methods. The more restrictive features of parametric methods are absent. No estimates of parameters or distributional assumptions are necessary. Measurements on a nominal or ordinal scale may be handled — it is in fact assumed that the data are in one or other of these forms, depending on the technique that is used. This is a strong point in favour of non-parametric methods, because many data in geography are not at interval or ratio level. Furthermore, interval or ratio measurements that fail to comply with the assumptions of a parametric procedure may be converted into ordinal or nominal measurements and analysed using a non-parametric procedure, and many of these procedures are capable of dealing with very small sample sizes. Finally, calculations may be more straightforward.

It may appear that all the advantages lie with non-parametric methods. However, the procedures as such do not give estimates of means and cannot readily provide confidence intervals. This is a drawback, especially in view of recent developments in statistical theory which allow confidence intervals to be provided for eccentrically distributed observations based on relatively small samples. Although more effort and more information must be put in when using parametric methods, the rewards consist of far more information coming out. Non-parametric methods are also less *power efficient*. Larger sample sizes are required in order to detect differences between means, compared with the equivalent parametric procedure. Specifically, there is a higher probability of committing a Type 2 error. Parametric methods usually repay the effort involved if they can be used. Non-parametric methods are extremely useful for testing hypotheses and may prove to be the only means of statistical analysis under certain circumstances.

In this chapter, non-parametric equivalents of the z and t tests for comparing two means, and the F test for comparing three or more means, will be introduced. These are based on ordinal data. Methods for comparing frequency distributions based on nominal data are also briefly discussed.

COMPARING TWO DISTRIBUTIONS – THE MANN–WHITNEY U TEST

The Mann–Whitney U test is the non-parametric equivalent of the parametric t and z tests, and is used to compare ordinal or ranked data obtained in random sampling from two populations, or from an equivalent natural sampling context.

Under the null hypothesis, the two populations are assumed to have the same shape and location (Fig. 13.1a). According to the alternative hypothesis, one of the two populations consists of measurements which, in general, are larger than those of the other. The test is one-tailed if the direction of the difference can be specified *a priori* (Fig. 13.1b); otherwise, it is two-tailed (Fig. 13.1c). If X_1 and X_2 represent randomly drawn observations from populations 1 and 2 respectively, then the hypotheses for a two-tailed test may be stated as

$$H_0 : P(X_1 > X_2) = 0 \cdot 5$$
$$H_1 : P(X_1 > X_2) \neq 0 \cdot 5$$

where $P(X_1 > X_2)$ is the probability that X_1 exceeds X_2. In words, H_0 states that the measurements in population 1 are in general no larger (or smaller) than those in population 2, or that the sample measurements were drawn from the same distribution and, by implication, have the same median. H_1 states that measurements in population 1 are in general different from those in population

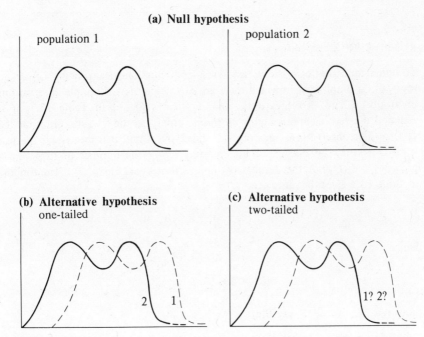

(a) Null hypothesis

population 1

population 2

(b) Alternative hypothesis
one-tailed

2   1

(c) Alternative hypothesis
two-tailed

1? 2?

Figure 13.1   Null and alternative hypotheses for the Mann–Whitney U test.

Table 13.1   All possible arrangements of ranked scores for $n_1 = 2$, $n_2 = 4$.

Ranked scores		Ranked scores denoting group	U_1	U_2	U
Sample 1	Sample 2				
1,2	3,4,5,6	A A B B B B	8	0	0
1,3	2,4,5,6	A B A B B B	7	1	1
1,4	2,3,5,6	A B B A B B	6	2	2
1,5	2,3,4,6	A B B B A B	5	3	3
1,6	2,3,4,5	A B B B B A	4	4	4
2,3	1,4,5,6	B A A B B B	6	2	2
2,4	1,3,5,6	B A B A B B	5	3	3
2,5	1,3,4,6	B A B B A B	4	4	4
2,6	1,3,4,5	B A B B B A	3	5	3
⋮	⋮	⋮	⋮	⋮	⋮

2, so that the medians are different. The hypotheses are couched in terms of medians rather than means because the test deals with ranks.

> State the appropriate hypotheses, in symbols and words, for a one-tailed test in which it is thought that measurements in population 2 are in general greater than those in population 1.

Sampling distribution of U

To understand the logic of the test, consider all possible arrangements of ranked scores for a simple case in which there are $n_1 = 2$ and $n_2 = 4$ sample observations in groups 1 and 2 respectively, and $n = (n_1 + n_2) = 6$. In Table 13.1, it is assumed that scores are ranked in *ascending* order, i.e. the lowest score is ranked 1. Denoting observations in groups 1 and 2 by A and B respectively, the test statistic U may be calculated by counting *either* the number of times an A precedes a B (U_1) *or* the number of times a B precedes an A (U_2). The top line in Table 13.1 gives the sequence

$$A \quad A \quad B \quad B \quad B \quad B$$

hence

$$U_1 = 2 + 2 + 2 + 2 = 8$$

since two A's precede every B, or

$$U_2 = 0 + 0 = 0$$

as neither A is preceded by a B. Conventionally, only the *smaller* of these two values is used, in this case $U = U_2 = 0$.

> Complete Table 13.1, also calculating the values of U_1 and U_2. Show that there are 15 possible arrangements, as given by the binomial coefficient $\binom{6}{2}$.

From the completed table, we can plot the *sampling distribution* of U_1 (Fig. 13.2). For example, $P(U_1 = 0) = 1/15 = 0.067$, as only one arrangement yields $U_1 = 0$. Given $n_1 = 2$, $n_2 = 4$, values of U_1 would occur with the probabilities specified in Figure 13.2 if the null hypothesis of a random allocation of ranks to groups were true.

> Show that $P(U_1 = 1) = 0.067$, $P(U_1 = 4) = 0.200$, $P(U_1 = 6)$ $= 0.133$. Given $n_1 = 2$, $n_2 = 4$, show that it is not possible to use the 5% significance level to reject H_0. Does this apply to both one- and two-tailed tests?

It is possible to construct such distributions for any values of n_1 and n_2, but this becomes tedious even for relatively small values. For $n_1 = 8$, $n_2 = 4$, there

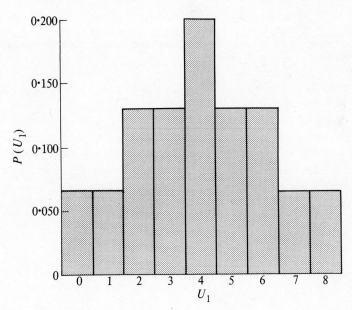

Figure 13.2   Sampling distribution of U_1 (number of A's preceding B's) for $n_1 = 2$, $n_2 = 4$.

are $\binom{12}{8} = 495$ possible arrangements, and for $n_1 = n_2 = 10$, we have $\binom{20}{10} = 184\ 756$.

Critical values of U have been tabulated as in Table A.6. Although the test may be used with samples of any size, the larger group should have a sample size $n_1 \geqslant 9$; if not, refer to Siegel (1956, pp. 117–19) or Hammond and McCullagh (1974, pp. 173–5). The group with the *larger* number of observations is known as group 2, this being the convention for tables of critical values. Assuming a significance level of $\alpha = 0.025$ for a one-tailed test and $\alpha = 0.05$ for a two-tailed test (this simple relationship holds because the sampling distribution of U is symmetrical – see Fig. 13.2), and $n_1 = 7, n_2 = 12$, the critical value of U given in Table A.6 is 18. The region of rejection includes the critical value and all values *below* it. Let us now consider an application of the test.

Examples

As part of a third-year undergraduate project (Bentall *et al* 1974), 50 randomly selected individuals from two areas in Reading were asked to identify 20 photographs of locations in the town. The photographs, ten of which were taken on main roads, and ten in nearby residential areas, were presented in a different random order to each person. As shown in Table 13.2, every site was given a score according to the number of respondents who identified it correctly. It was hypothesised in advance that knowledge of sites on main roads (group 1) would in general be greater than that of sites in residential areas (group 2). The statistical hypotheses are therefore

$H_0: P(X_1 > X_2) = 0.5$ (the 'road median' and 'area median' do not differ)

$H_1: P(X_1 > X_2) > 0.5$ (the 'road median' exceeds the 'area median')

Table 13.2   'Roads' and 'areas' scores and ranks.

	Group 1 (roads)			Group 2 (areas)	
	Score	Rank		Score	Rank
	43	19·5		13	5·5
	27	13		24	11
	13	5·5		23	9·5
	33	16		27	13
	43	19·5		9	1·5
	42	18		9	1·5
	35	17		10	3
	28	15		23	9·5
	22	8		12	4
	27	13		17	7
Median	30·5	$R_1 = 144.5$	Median	15·0	$R_2 = 65.5$

We can proceed with the one-tailed test because the sample median for roads is 30·5, and that for areas, 15·0 (Table 13.2). Ranks are assigned to the scores, *ignoring group membership*, the lowest score receiving rank 1, the next lowest rank 2, etc. For tied scores, the rank allocated is the mean of the ranks that would have been given to untied scores. For example, there are two sites with the lowest score of 9, so each receives the rank $(1 + 2)/2 = 1·5$. To ease calculation, U may be found from

$$U = n_1 n_2 + \tfrac{1}{2}n_1(n_1 + 1) - R_1 \tag{13.1}$$

or

$$U = n_1 n_2 + \tfrac{1}{2}n_2(n_2 + 1) - R_2 \tag{13.2}$$

where R_1 and R_2 are the sums of the ranks given to the n_1 road values and n_2 area values, respectively (Table 13.2). The *smaller* value of U must be used. (The value of U obtained from these formulae may differ from that obtained by counting, because the formulae allow for tied ranks involving items from *both* groups. It is wiser to use the formulae, as they allow for all possible orderings within ties.)

Use (13.1) and (13.2) to show that $U = 10·5$, and Table A.6 to show that U is significant at the 0·025 and 0·01 levels.

Because Table A.6 does not cater for groups in which the sample size exceeds 20, a different form of the test is used for 'large' samples. It has been found that as n_1 and n_2 increase, the sampling distribution of U rapidly approaches a normal distribution with mean

$$\mu_U = n_1 n_2 / 2 \tag{13.3}$$

and standard error

$$\sigma_U = \sqrt{\frac{n_1 n_2 (n_1 + n_2 + 1)}{12}} \tag{13.4}$$

so that the test statistic is the standard normal deviate

$$z = \frac{U - \mu_U}{\sigma_U} \tag{13.5}$$

whose critical values are given in Table A.3. Provided the sample medians are checked, it does not matter whether (13.1) or (13.2) is used to determine U, as its sampling distribution is symmetrical.

Of the 50 individuals interviewed in the Reading knowledge study, 25 were from an area of 'high' socio-economic status (HES) and 25 from an area of 'low socio-economic status' (LES). Both areas were within a mile of each other and roughly equidistant from the town centre, in an attempt to control for the possible confounded effects of status and location. Scores for the individuals in each group are shown separately for road and area sites in Table 13.3. A non-parametric test was used because a plot of the HES scores shows evidence of bimodality, and the LES scores are left-skewed. It was hypothesised that the HES group would generally show greater knowledge of road sites than the LES group, because higher income people tend to be more mobile.

State the statistical hypotheses, in symbols and words, letting group 1 represent LES, and group 2 HES. Should we go ahead with the test, given the sample median values of Table 13.3?

Table 13.3   Recognition of roads and areas by socio-economic groups.

Low socio-economic status		High socio-economic status	
Number correct out of 10 for		Number correct out of 10 for	
Roads	Areas	Roads	Areas
8 (40)*	4	6 (23)	5
5 (17)	4	7 (31·5)	8
9 (45)	5	2 (1·5)	2
4 (10·5)	0	7 (31·5)	3
3 (4·5)	3	10 (48·5)	6
5 (17)	1	7 (31·5)	2
2 (1·5)	3	10 (48·5)	6
4 (10·5)	3	10 (48·5)	6
10 (48·5)	7	8 (40)	6
3 (4·5)	2	9 (45)	5
8 (40)	5	8 (40)	4
4 (10·5)	2	8 (40)	4
7 (31·5)	4	7 (31·5)	3
9 (45)	2	6 (23)	3
7 (31·5)	4	7 (31·5)	2
7 (31·5)	5	5 (17)	4
6 (23)	0	6 (23)	3
4 (10·5)	1	7 (31·5)	2
3 (4·5)	1	6 (23)	3
6 (23)	3	4 (10·5)	4
4 (10·5)	4	4 (10·5)	3
7 (31·5)	3	5 (17)	2
3 (4·5)	3	4 (10·5)	3
8 (40)	0	8 (40)	4
6 (23)	1	5 (17)	1
$R_1 = 559\cdot5$		$R_2 = 715\cdot5$	
Median = 5·5		Median = 7·0	

*Ranks for road scores are given in brackets.

Sample sizes are again equal since $n_1 = n_2 = 25$ and, from Table 13.3, the sums of ranks are $R_1 = 559 \cdot 5$ and $R_2 = 715 \cdot 5$. By (13.1), we have

$$U = (25 \times 25) + \tfrac{1}{2}(25 \times 26) - 559 \cdot 5 = 390 \cdot 5$$

$$- 715 \cdot 5 = 234 \cdot 5$$

and by (13.3) and (13.4), we have

$$\mu_U = \frac{25 \times 25}{2} = 312 \cdot 5$$

and

$$\sigma_U = \sqrt{\frac{(25 \times 25) \times 51}{12}} = \sqrt{\frac{31875}{12}} = 51 \cdot 54$$

Show that $z = 1 \cdot 51$. Can H_0 be rejected at the 5% significance level?

Correction for ties

If ties occur between observations in the same group, neither U nor σ_U is affected. However, a correction is necessary in the large-sample case, otherwise σ_U tends to be overestimated. Corrected for ties, the standard error is

$$\sigma_U = \sqrt{\left(\frac{n_1 n_2}{n(n-1)}\right) \times \left(\frac{n^3 - n}{12} - \Sigma T\right)} \qquad (13.6)$$

where $n = (n_1 + n_2)$ as before, and

$$T = \frac{t^3 - t}{12}$$

t being the number of observations tied at the same rank. Tied scores, and their frequencies, from Table 13.3 are listed below (Table 13.4). Although four observations are tied with a rank of $4 \cdot 5$, they are all in the LES group and so are ignored. It is found that $T = 198 \cdot 0$. Substituting in (13.6) gives

$$\sigma_U = \sqrt{\left(\frac{25 \times 25}{50 \times 49}\right) \times \left(\frac{50^3 - 50}{12} - 198\right)}$$

$$= \sqrt{0 \cdot 255 \times 10214 \cdot 5} = 51 \cdot 04$$

and

$$z = \frac{390 \cdot 5 - 312 \cdot 5}{51 \cdot 04} = 1 \cdot 53$$

This value of z is slightly higher than that originally obtained. Even where the proportion of ties is large (92% of the observations were involved in intergroup ties in this example), the correction factor is relatively small. The correction is only really worthwhile if there are many ties of a high frequency — compare $(t^3 - t)/12$ for t values of 2 and 7 in Table 13.4 — and if the value of z based on an uncorrected σ_U lies close to a critical significance level.

Table 13.4   Calculation of correction for ties for socio-economic group road scores.

Score	Rank	Frequency (t)	$t^3 - t$	$T = (t^3 - t)/12$
2	1·5	2	6	0·5
4	10·5	8	504	42·0
5	17·0	5	120	10·0
6	23·0	7	336	28·0
7	31·5	10	990	82·5
8	40·0	7	336	28·0
9	45·0	3	24	2·0
10	48·0	4	60	5·0
–	–	–	–	$\Sigma T = 198 \cdot 0$

COMPARING THREE OR MORE DISTRIBUTIONS – THE KRUSKAL–WALLIS ANALYSIS OF VARIANCE BY RANKS

The Kruskal–Wallis test provides a non-parametric or distribution-free equivalent to the parametric F test for the Analysis of Variance. It is assumed that ordinal or ranked data are available in random sampling from three or more populations, or from a natural sampling context.

Assuming there are $k = 3$ populations or groups, the null hypothesis holds that there is no difference between the three population distributions. Suppose we had sample knowledge scores with respect to main road sites (group 1), residential area sites (group 2) and town centre sites (group 3), and X_1, X_2 and X_3 represent typical sample observations for these groups. Then the null hypothesis is formally stated as

$$H_0 : P(X_1 > X_2) = P(X_1 > X_3) = P(X_2 > X_3) = 0 \cdot 5$$

or, in words, a sample score for any one group or population of sites is just as likely to exceed (or fall below) a sample score for any other group or population of sites. The alternative hypothesis states that at least one inequality occurs, i.e.

H_1: for at least one pair of site types, i and j, $P(X_i > X_j) \neq 0.5$

Normally, we simply state

H_0 : all the median scores are equal

H_1 : at least two of the median scores are unequal

If there were four sites in each group, and observations are ranked by assigning rank 1 to the lowest score, rank 2 to the next lowest, etc, and the sample scores were randomly drawn from identical distributions with identical medians, the sums of the ranks should be more or less equal, as shown in Table 13.5a. For non-identical distributions and medians, the sums of the ranks are likely to be unequal, as shown in Table 13.5b. An index of the difference between the sums of ranks is provided by the test statistic

$$H = \frac{12}{n(n+1)} \times \left(\sum_{i=1}^{k} \frac{R_i^2}{n_i} \right) \qquad (13.7)$$

Table 13.5 Hypothetical observations for Kruskal–Wallis test (S = score, R = rank).

(a) Identical distributions and medians.

Group 1		Group 2		Group 3	
S	R	S	R	S	R
16	8	9	1	12	3
13	4	15	6·5	20	11
17	9	21	12	18	10
14	5	15	6·5	10	2
$R_1 = 26$		$R_2 = 26$		$R_3 = 26$	
Medians 6·5		6·5		6·5	

(b) Non-identical distributions and medians.

Group 1		Group 2		Group 3	
S	R	S	R	S	R
16	8	9	1	21	12
18	10	15	6·5	15	6·5
17	9	10	2	14	5
20	11	12	3	13	4
$R_1 = 38$		$R_2 = 12·5$		$R_3 = 27·5$	
Medians 9·5		4·75		5·75	

where R_i is the sum of the sample ranks, n_i is the sample size in the ith group, and n is the total number of observations.

> Calculate H first with respect to the hypothetical observations given in Table 13.5a, and then for those in Table 13.5b. Does H appear to provide a suitable index of the difference between groups?

If there are three *small samples*, i.e. the number of observations in each does not exceed 5, special tables must be used to test the significance of H. For an exposition and examples, see Siegel (1956, pp. 184–93) or Hammond and McCullagh (1974, pp. 186–91).

Provided there are three or more samples, and at least 5 observations in each, a test for *large samples* may be employed. Under H_0, the value of H follows a χ^2 distribution with $(k - 1)$ degrees of freedom, where k is the number of groups. A *one-tailed* test is appropriate.

In the study previously described (pp. 170–1), Bowlby also investigated the degree to which individuals thought they used the sample stores. Perceived use

Table 13.6   Observations on perceived store use for the three sectors defined by Bowlby.

North		West		South	
Score	Rank	Score	Rank	Score	Rank
19·8	48	21·1	50	19·9	49
16·6	38	19·7	47	16·4	36
15·5	30	14·8	26	3·1	9
17·0	40	19·4	45	9·3	19·5
12·6	23	3·1	9	3·1	9
16·7	39	17·1	41	14·7	25
18·8	44	16·6	37	3·1	9
19·5	46	15·9	32	3·1	9
14·8	27	7·2	18	3·1	9
17·6	43	3·1	9	3·1	9
16·0	33	3·1	9	16·3	35
16·1	34	3·1	9	3·1	9
3·1	9			15·6	31
13·8	24			12·3	22
15·3	29			11·4	21
15·2	28			9·3	19·5
				3·1	9
				3·1	9
				3·1	9
				3·1	9
				3·1	9
				17·4	42
$R_1 = 535$		$R_2 = 332$		$R_3 = 408$	
$n_1 = 16$		$n_2 = 12$		$n_3 = 22$	

scores are given in Table 13.6 for stores in north, west and south sectors radiating from the place of residence. The number of stores in the north sector was $n_1 = 16$, in the west sector $n_2 = 12$, and in the south sector $n_3 = 22$. Thus $n = n_1 + n_2 + n_3 = 50$. Histograms of the observations suggest strong skewness, in different directions, for the north and south sectors, and bimodality for the west sector. A non-parametric test seems appropriate.

The hypotheses to be tested are

H_0: the median perceived use scores for the stores in the three sectors are equal

H_1: at least two of the median use scores are unequal

As before, we assign ranks to the perceived use scores shown in Table 13.6, the sums of ranks being $R_1 = 535$ (north), $R_2 = 332$ (west) and $R_3 = 408$ (south). A check shows $\Sigma R_i = 1275$, as it should, since $n \times (n + 1)/2 = (50 \times 51)/2 = 1275$. Substituting in (13.7) gives

$$H = \left(\frac{12}{50 \times 51}\right) \times \left(\frac{535^2}{16} + \frac{332^2}{12} + \frac{408^2}{22}\right) - [3 \times (50 + 1)]$$

$$= (0{\cdot}0047 \times 34641) - 153 = 162{\cdot}81 - 153 = 9{\cdot}81$$

This test statistic H has $(k - 1)$ or $(3 - 1) = 2$ degrees of freedom. Reference to Table A.1 shows that, for a one-tailed test, $\chi_2^2 = 5{\cdot}99$ for $\alpha = 0{\cdot}05$, or $\chi_2^2 = 9{\cdot}21$ for $\alpha = 0{\cdot}01$. We conclude that a difference between at least two of the sector means exists at the 1% significance level.

The effect of ties is to diminish the value of H, and we may divide H by

$$1 - \frac{12\Sigma T}{n^3 - n} \tag{13.8}$$

where $T = (t^3 - t)/12$. Table 13.6 shows only one set of tied ranks occurring in more than one group — there are 17 scores of $3{\cdot}1$ given a rank of 9. We thus obtain $(t^3 - t)/12 = (17^3 - 17)/12 = 408$ which, substituted in (13.8), gives

$$1 - \frac{12 \times 408}{50^3 - 50} = 1 - 0{\cdot}039 = 0{\cdot}961$$

Dividing H by this value gives

$$\chi_2^2 = \frac{H}{1 - \Sigma T/(n^3 - n)} = \frac{9{\cdot}81}{0{\cdot}961} = 10{\cdot}21$$

a value differing relatively little from the original.

COMPARISON OF DISTRIBUTIONS BASED ON NOMINAL OR CATEGORICAL DATA

The χ^2 test, whose major features were discussed in a natural sampling context in Chapter 5, should be used to compare distributions based on **nominal** or **categorical** data. In an artificial sampling context, the investigator wishes to decide whether the observations have been obtained in random sampling from a population in which observations are independently assigned to categories.

The **one-sample test**, permitting comparison of a single observed frequency distribution with an expected frequency distribution, may be regarded as a test for differences between two or more *proportions* based on the hypotheses

$$H_0: \ \Pi_1 = \Pi_2 = \ldots = \Pi_k = \Pi$$

$$H_1: \text{ at least two of the } \Pi_i \text{ are different}$$

where k is the number of categories compared. Calculations are as described in the first part of Chapter 5 (pp. 46–8).

The R-sample test, where $R \geqslant 2$, is for differences between two or more *distributions*, i.e. if F_1 represents the frequency distribution of observations in population 1, F_2 that of observations in population 2, etc, then we test the hypotheses

$$H_0: \ F_1 = F_2 = \ldots = F_R = F$$

$$H_1: \text{ at least two of the } F_i \text{ are different}$$

where R is the number of distributions compared. Calculations are as detailed for a χ^2 test used as a measure of association (pp. 59–61).

EXERCISES

1. Use the scores given in Table 13.3 for area sites to compare the higher and lower socio-economic groups. Do histograms of the observations seem to justify a non-parametric test? Given $s_1^2 = 5 \cdot 06$, $s_2^2 = 4 \cdot 16$ and $n_1 = n_2 = 25$, do you think use of the Mann–Whitney test can be justified on the grounds of unequal variances? Whatever your decision, carry out the Mann–Whitney test on the observations, and carefully justify the choice of a one- or two-tailed test. Also calculate the value of z corrected for ties.

2. Doornkamp and King (1971, p. 41) wished to ascertain whether drainage basin areas in the Kyogya Valley and Lake Nagugabo, two distinct morphological regions in their study area in Uganda, differed only because of chance variations or not. The observations, ranked for each area, are given below (in square miles).

Kyogya	4·08	4·53	4·92	6·18	6·22	6·54	7·55	8·92	10·16	12·50
Nagugabo	2·11	2·18	2·26	2·47	2·60	3·78	4·56	4·85	6·62	

According to Doornkamp and King, the basin areas in the two regions have differing variances $s_1^2 = 2 \cdot 50$ (Kyogya) and $s_2^2 = 7 \cdot 17$. Is this sufficient justification for using the Mann–Whitney test? Are there any properties of the observations which make them unsuitable for a *t*-test? Carry out the Mann—Whitney test if you think it justified.

3. Using 18 categories for orientation (in degrees clockwise from north) and four categories for dip (from 0 to 90° in the direction of slope of the underlying surface), Andrews (1963) was able to classify 200 stones from a cross-valley moraine as shown below.

Orientation		Dip	
Category	Frequency	Category	Frequency
0–19	7	0–19	151
20–39	6	20–39	33
40–59	2	40–59	13
60–79	1	60–79	3
80–99	10		Σ 200
100–19	15		
120–39	5		
140–59	4	(Andrews excluded the	
160–79	8	80–90° class as it	
180–99	44	contained no observations.)	
200–19	20		
220–39	7		
240–59	3		
260–79	36		
280–99	19		
300–19	4		
320–39	5		
340–59	4		
Σ 200			

(handwritten: e = 50)

(handwritten: e = 11.11)

(a) Calculate the 'orientation strength' given by the χ^2 statistic for the orientation measurements. Then, calculate 'dip strength' in the same way for the dip measurements.

(b) What do tests on the χ^2 statistics against the null hypothesis of independent and randomly distributed observations over the categories show?

4. In a study of class differences in environmental perception, Goodchild (1974) asked 22 middle-class and 27 working-class individuals, all randomly selected, to rate a number of factors influencing their housing preferences as either very important (V), fairly important (F), not very important (N) or unimportant (U). Some of the responses are summarised below.

	Importance							
	Middle-class				Working-class			
Item	V	F	N	U	V	F	N	U
Near family	1	13	7	1	5	8	12	2
Plenty of work	12	3	6	1	16	8	3	0
Kind of house I want	10	10	2	0	18	6	3	0
Near good school	15	3	4	0	19	6	2	0
Near shops	3	13	6	0	13	12	2	0
Attractive surroundings	14	7	1	0	8	18	1	0
Not too expensive	8	6	8	0	14	12	1	0

(various handwritten annotations throughout the table)

(handwritten at bottom: 27/50 22/50   85/27   22 9 30/3)

For which items is there a significant difference between the responses given by the two groups? Why should you be very careful in checking one of the assumptions underlying use of the test?

5. Carey and Mapes (1972, p. 90) examined the relationship between functional distance (defined as the distance that a person would need to walk from the dwelling to the adjacent one using the quickest conventional route) and number of visits, as summarised in the table.

Number of visits	Units of functional distance			Totals
	1	2	$\geqslant 3$	
1	80	45	72	197
2	26	4	11	41
$\geqslant 3$	64	16	31	111
Totals	170	65	114	349

Is there a significant association between number of visits and functional distance? Is it positive or negative?

REFERENCES AND FURTHER READING

A wide variety of non-parametric techniques is discussed in:

Conover, W. J. 1971. *Practical nonparametric statistics.* New York: Wiley.

Mosteller, F. E. and R. E. K. Rourke 1973. *Sturdy statistics: nonparametric and order statistics.* Reading, Mass.: Addison-Wesley.

Siegel, S. 1956. *Nonparametric statistics for the behavioral scientist.* New York: McGraw-Hill.

The use of such techniques in geography has been strongly advocated in:

French, H. M. 1971. Quantitative methods and non-parametric statistics. In *Quantitative and qualitative geography – nécessité d'un dialogue,* H. M. French and J. B. Racine (eds). 119–28. University of Ottawa, Dept of Geography, Occasional Papers no. 1.

For a book that covers a wide variety of techniques, together with worked examples and applications to real-world data, see:

Lewis, P., 1977. *Maps and statistics.* London: Methuen.

In physical geography, numerous examples are to be found in:

Doornkamp, J. C. and C. A. M. King 1971. *Numerical analysis in geomorphology.* London: Edward Arnold.

For individual studies, see:

Andrews, J. T. 1963. The cross-valley moraines of north-central Baffin Island: a quantitative analysis. *Geog. Bull.* **20,** 82–129.

Driscoll, E. M. and D. Hopley 1968. Coastal development in a part of tropical Queensland, Australia. *J. Trop. Geog.* **26,** 17–28.

Fenwick, I. M. 1965. Some problems of soil permeability measurements. In *Essays in geomorphology for Austin Miller,* J. B. Whittow and P. D. Wood (eds), 168–87. Reading: University of Reading.

For human geographers, there is a variety of papers on social and behavioural topics, nearly all based on the χ^2 test, such as:

Carey, L. and R. Mapes 1972. *The sociology of planning.* London: Batsford.

Davidson, R. N. 1976. Social deprivation: an analysis of intercensal change. *Trans Inst. Br. Geogs* (New Series) 1((1), 108–17.

Goodchild, B. 1974. Class differences in environmental perception: an exploratory study. *Urban Stud*. 11(2), 157–69.

Herbert, D. T. 1976. The study of delinquency areas: a social–geographical approach. *Trans Inst. Br. Geogs* (New Series) 1(4), 472–92.

Pocock, D. C. D. 1976. Some characteristics of mental maps: an empirical study. *Trans Inst. Br. Geogs* (New Series) 1(4), 493–512.

Toomey, D. M. 1970. The importance of social networks in working-class areas. *Urban Stud*. 7(3), 259–70.

Further examples of the use of the Mann–Whitney and Kruskal–Wallis tests are given in Lewis (1977, pp. 152–6) as referenced above, and in:

Hammond, R. and P. S. McCullagh 1974. *Quantitative techniques in geography*, Oxford: Oxford University Press. pp. 172–80, 186–91).

Data for the knowledge study are from:

Bentall, V. A., S. J. A. Fell, R. Lee and J. Sorsby 1974. *Subjective knowledge levels in Reading*. University of Reading, Dept of Geography, Unpublished Behavioural Geography Project.

14 Simple Correlation

INTRODUCTION

Simple correlation techniques deal with associations or relationships between two variables providing measures of the **strength** of the association and statistical tests of its **significance**. In the first part of this chapter we deal with two non-parametric coefficients of correlation — the phi (ϕ) coefficient and Spearman's rank correlation (r_S) coefficient — and, in the second, with a widely used parametric technique — Pearson's product-moment correlation coefficient.

NON-PARAMETRIC TECHNIQUES

Use of non-parametric correlation techniques is advocated, first, when true nominal or ordinal data are available, e.g. two groups are asked to rank areas in order of residential desirability; secondly, when interval or ratio data are available but deviate markedly from normality and transformations prove ineffective; and thirdly, when interval or ratio measurements are thought to be rather imprecise.

The **phi coefficient**, ϕ, is used to determine the strength of the association between two *binary* variables, i.e. nominal variables with two categories. As we saw in Chapter 5, the χ^2 statistic provides a test of the significance of the association between such variables. If the data are set out in a 2×2 contingency table as shown in Table 14.1, the value of ϕ is given by

$$\phi = \frac{AD - BC}{\sqrt{(A + B)(C + D)(A + C)(B + D)}} \tag{14.1}$$

Table 14.1   Contingency table with quantities for calculating phi (ϕ) coefficient.

Variable 2	Variable 1		
	Category 1	Category 2	Totals
Category 1	A	B	$A + B$
Category 2	C	D	$C + D$
Totals	$A + C$	$B + D$	n

Table 14.2   Rank orders for different experimental groups for recall of items seen from the road.

Item		Passengers	Commuters	Drivers	d_i	d_i^2
1	Skyline	1	4	6	2	4
2	Houses (early)	2	2	3·5	1·5	2·25
3	Overpass (early)	22	8	9	1	1
4	Mystic River Bridge	3	3	1	2	4
5	Billboards (end)	7	14	3·5	10·5	110·25
6	Industry (late)	5	7	8	1	1
7	Charlestown	17	20	13	7	49
8	Government Centre	6	1	16	15	225
9	Overpass (end)	12·5	23	5	18	324
10	Industry (early)	4	9	10·5	1·5	2·25
11	Chelsea City Hall	16	25	13	12	144
12	Custom House Tower	14	13	13	0	0
13	Naval Hospital	23	15	23·5	8·5	72·25
14	Billboards (early)	12·5	25	7	18	324
15	Charles River Park	19	21	23·5	2·5	6·25
16	State Street Park	25·5	16	23·5	7·5	56·25
17	Toll Booth	8	11	2	9	81
18	Soldiers' Home	21	12	10·5	1·5	2·25
19	Twin-Tower Church	15	22	18	4	16
20	Bunker Hill Monument	11	5	23·5	18·5	342·25
21	Prudential	10	6	19	13	169
22	Three-Deckers	9	10	15	5	25
23	Smokestacks	24	19	17	2	4
24	150 Causeway Building	25·5	26	23·5	2·5	6·25
25	John Hancock	20	18	20	2	4
26	Sign: Callan Tunnel	18	17	23·5	6·5	42·25
						$\Sigma 2017\cdot50$

and ranges in value from +1 for a perfect *positive* association, to −1 for a perfect *negative* association between the two variables. If there is *no* association, $\phi = 0$.

> Show this to be so for the values of ϕ obtained for the associations between plant species A and B in Tables 5.4a, 5.4b and 5.4e.

Spearman's rank correlation coefficient, r_S, may be calculated if the n observations on each of the two variables, X and Y, thought to be associated, are ranked separately from 1 to n, substituting average ranks for tied ranks where appropriate. For example, Carr and Schissler (1969) asked a number of different groups of individuals to recall items they had seen immediately after travelling along an expressway into downtown Boston. Average rankings for those individuals who were regular highway commuters by that route into Boston but who otherwise took no part in the experiment (X values) were compared with those who participated in the experiment as drivers (Y values) (Table 14.2). If the ranks calculated for the commuters' and drivers' responses were the same with respect to each of the $n = 26$ items in Table 14.2, then

$$d_i = X_i - Y_i$$

would equal zero for each item. Any disparity between the two sets of rankings would produce a non-zero value of d_i. The greater the disparity, the larger the value of d_i, provided the sign of d_i is ignored. To ensure that negative and positive differences do not cancel out, a measure of correlation based on d_i^2 is used. The formula for the rank correlation coefficient is

$$r_S = 1 - \frac{6 \, \Sigma d_i^2}{n^3 - n} \tag{14.2}$$

For the derivation of this formula, see Siegel (1956, pp. 203–4). Elements of Σd_i^2 are given in Table 14.2, and we find $\Sigma d_i^2 = 2017 \cdot 5$.

Substitute this value, and $n = 26$, in (14.2) and show that $r_S = 0 \cdot 31$.

Like the ϕ coefficient, r_S also ranges between $-1 \cdot 0$ and $+1 \cdot 0$, the former representing a perfect negative association, the latter a perfect positive association, between the two sets of ranks. In this case, the association between the items recalled by members of the two groups is only weakly positive.

The basic formula does not allow for the effects of tied ranks. However, the proportion of such ranks must be rather large to have any appreciable influence and so the corrected formula will not be considered here; for details, see Siegel (1956, pp. 206–10).

We can also test whether the observed rank correlation is generated purely by chance, i.e. is the value of r_S likely to have been randomly generated from a population, or by a process, for which the parameter ρ_S (rho subscript S) equals zero? For a two-tailed test, we define the statistical hypotheses

$$H_0: \ \rho_S = 0$$

$$H_1: \ \rho_S \neq 0$$

If the number of pairs of observations is less than $n = 10$, a special table must be used to determine the critical values, and hence the regions of rejection, for r_S, for $\alpha = 0 \cdot 05$ or $0 \cdot 01$. Such cases are rare in geography, and so the reader is referred to Siegel (1956, pp. 210–12). For $n \geqslant 10$, it may be shown that the sampling distribution of r_S follows a t distribution with $(n - 2)$ degrees of freedom when H_0 is true, with expected value $\rho_S = 0$ and estimated standard error

$$s(r_S) = \sqrt{(1 - r_S^2)/(n - 2)}$$

yielding the test statistic

$$t_{n-2} = r_S \sqrt{\frac{n-2}{1-r_S^2}} \qquad (14.3)$$

whose critical values are given in Table A.4.

> Show that the test statistic with respect to the commuters and drivers of Table 14.2 is 1·60. Is there sufficient evidence to reject H_0?

The degree of association between three or more sets of ranks may be determined by Kendall's **coefficient of concordance**, W, and its significance tested using a χ^2 test. Details may be found in Siegel (1956, pp. 229–38).

PARAMETRIC TECHNIQUES

These techniques deal with variables measured on an interval or ratio scale, and further assumptions are necessary if tests of significance are to be carried out.

In Figure 14.1a, the relationship shown in the scatter diagram between observations on rainfall (X) on the horizontal axis and run-off (Y) on the vertical axis is *positive* because the larger the values of one variable, the larger the associated values of the other. Conventionally, Y is the variable whose behaviour particularly interests us and which we should like to explain. In Figure 14.1b the association between familiarity with a retail store (Y) and distance to the store (X) is *negative* because, as we might expect, the larger the values of X, the smaller the values of Y. The observations corresponding to these points are listed in Table 14.3. Schumm (1956) plotted depth of erosion (X) against per cent distance from the top of straight slope segments (Y) and found *no relationship* (Fig. 14.1c), i.e. values of Y seem neither to increase nor decrease with values of X.

The strength of the relationship between two interval or ratio variables, X and Y, is measured by

$$r = \frac{\Sigma(X_i - \bar{X})(Y_i - \bar{Y})}{\sqrt{[\Sigma(X_i - \bar{X})^2][\Sigma(Y_i - \bar{Y})^2]}} \qquad (14.4)$$

where r is the (product-moment) **correlation coefficient**, and n the number of pairs of observations. The term 'product-moment' is used because each of the terms in the numerator of Formula 14.4 represents an element of the first moment about a mean, i.e. $(X_i - \bar{X})$ is an element of the moment $\Sigma(X_i - \bar{X})$, and similarly for $(Y_i - \bar{Y})$. The numerator can therefore be regarded as a product-moment. Because the denominator of (14.4) must always be positive

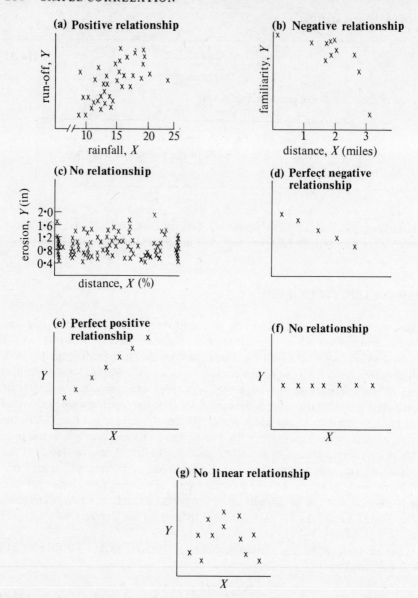

Figure 14.1   Examples of relationships between variables measured on interval or ratio scale.

(why?), the sign of r depends upon the sign of the product-moment term. Figure 14.2 shows that if the relationship between X and Y is positive, the majority of terms in $\Sigma(X_i - \bar{X})(Y_i - \bar{Y})$ will be *positive*, as we are multiplying two positive terms together in the 'north-east' quadrant and two negative terms in the 'south-west' quadrant; hence $r > 0$.

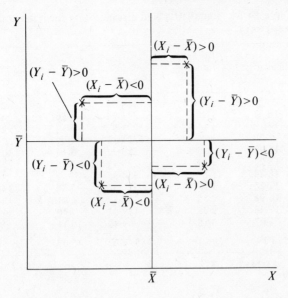

Figure 14.2   Terms making up the denominator of the product–moment correlation coefficient, r.

> Provide a logical argument to show that $r < 0$ if the relationship is negative, and $r \simeq 0$ if there is no relationship.

For a perfect negative relationship between X and Y, in which all observations lie on the same straight line (Fig. 14.1d), $r = -1 \cdot 0$; for a perfect positive relationship, $r = +1 \cdot 0$ (Fig. 14.1e); if there is no relationship, $r = 0 \cdot 0$. There are several ways in which a value of $r = 0 \cdot 0$ might occur: the observations may be more or less randomly scattered as in Figure 14.1c, lie on the same horizontal line (Fig. 14.1f) or show a strong but non-linear relationship (Fig. 14.1g).

Notice that it does not matter which axis we use for each variable in any of the examples in Figure 14.1, because r is symmetric, i.e. the correlation between X and Y gives the same result as that between Y and X.

> Inspect Formula 14.4 and convince yourself that interchanging terms in X_i and Y_i makes no difference.

To calculate r, use the computational formula

$$r = \frac{\Sigma X_i Y_i - (\Sigma X_i)(\Sigma Y_i)/n}{\sqrt{\Sigma X_i^2 - (\Sigma X_i)^2/n} \, \sqrt{\Sigma Y_i^2 - (\Sigma Y_i)^2/n}}$$
(14.5)

Table 14.3   Familiarity and distance data for Evanston area stores.

Familiarity Y_i	Distance (miles) X_i	Y_i^2	X_i^2	X_iY_i
20·85	0·14	434·72	0·02	2·92
18·47	1·25	341·14	1·56	23·09
14·18	1·61	201·07	2·59	22·83
17·92	1·70	321·13	2·89	30·46
16·98	1·77	288·32	3·13	30·05
15·27	1·81	233·17	3·28	27·64
18·84	1·87	354·95	3·50	35·23
17·25	1·97	297·56	3·88	33·98
16·98	2·59	288·32	6·71	43·98
13·38	2·72	179·02	7·40	36·39
3·07	3·18	9·42	10·11	9·76
Σ173·19	20·61	2948·82	45·07	296·30

$\bar{Y} = 15\cdot74$, $\bar{X} = 1\cdot87$.

where each item is represented by a pair of observations, (X_i, Y_i), and n is the number of pairs of observations. In Bowlby's study of the relationship between familiarity and distance (Fig. 14.1b), the first store in Table 14.3 lies $X_1 = 0\cdot14$ miles from the individual's home area, and receives a familiarity score of $Y_1 = 20\cdot85$. Using the totals from this table, which are for stores lying in a sector north of the study area, gives

$$r = \frac{296\cdot30 - (173\cdot19)(20\cdot61)/11}{\sqrt{[45\cdot07 - (20\cdot61)^2/11]}\ \sqrt{[2948\cdot82 - (173\cdot19)^2/11]}} = \frac{-28\cdot20}{37\cdot85} = -0\cdot745$$

showing a moderately strong negative linear relationship.

INTERPRETING THE RESULTS OF A CORRELATION ANALYSIS

Once a statistical correlation has been established, the result may be interpreted in a number of ways. It may seem reasonable to believe that one variable, X, *causes* changes in the other, Y, as in the relationship between rainfall and run-off of Figure 14.1a, or between number of square feet of office space and number of trips generated from a building. This type of situation need not mean that only one causal factor is operating, however, and the scatter of observations about a perfect straight line in Figure 14.1a may indicate the influence of other, possibly less important, systematic factors. A second possibility is the *mutual interaction* of two variables, examples of which are common in geomorphology. Relationships between various measures of stream channel geometry, and in turn between these factors and valley-side slope characteristics, are taken to indicate a fine adjustment of various forms, one to another, and to the processes at work.

Carey and Mapes' study (1972) of social activity on new housing estates established relationships between the amount of estate visiting and variables such as age, number of visits made to relatives and number of homes previously occupied. It is more realistic to talk of many of these factors 'varying together', rather than in terms of one-way causation. Thirdly, a relatively strong linear association between two variables may be *spurious* because both reflect the influence of a third, controlling, variable. For example, Smith (1976) carried out a study of spatial information fields of consumers in Hamilton, Ontario, and asked the adult member of the household who most frequently purchases groceries to write down the locations of known grocery stores within the city. The correlation between number of reported shops and length of residence in Hamilton was found to be 0·685. However, it is quite possible that individuals of higher social status tend to have lived longer in Hamilton than individuals of lower social status. As such individuals generally have higher status and greater local mobility, it could well be that higher status is the truly important controlling factor. It is also possible that individuals of higher social status, being generally better educated, can more successfully cope with the task of writing down names of stores, so that ability in this respect, as much as spatial knowledge, is being measured. Fourthly, care should be taken when the results of a correlation are contrary to reasonable expectation. In a study of the relationships between accidents involving serious casualties and various other factors for 90 regions in England, Scotland and Wales, Meadows (1977) found a negative correlation of $r = -0·239$ between number of accidents and number of days in the year with precipitation. The linear association is weak, but runs contrary to expectation. Surely wet roads mean more dangerous driving con-

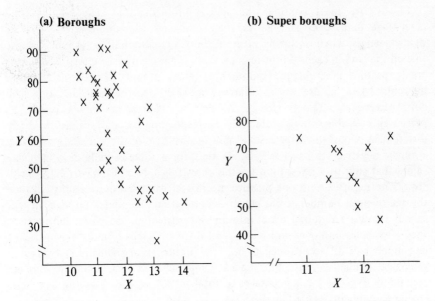

Figure 14.3   Effects of change in areal units on correlation coefficients.

ditions, and hence more accidents? However, it may be possible to explain such a result if we remember that more drivers tend to be attracted onto the road in good weather so that the *absolute* number of accidents occurring increases under these conditions, even though the *relative proportion* may decrease.

As indicated earlier (Ch. 5), the problem of ecological correlation may also arise because geographers often deal with data in which individual persons, households or other units have been aggregated into administrative regions. Relationships holding at one level of analysis may bear little resemblance to those revealed at another. Taking the percentage of households with exclusive use of washing and toilet facilities as a crude indicator of housing conditions, we should expect a negative correlation with adjusted mortality rate with respect to the 33 London boroughs, and obtained $r = -0.72$. However, if the 33 boroughs are combined to give 11 'super boroughs', we find $r = -0.23$. Figures 14.3a and 14.3b show scatter diagrams of observations for the two cases. Because we have combined contiguous boroughs with high and low mortality rates, the amount of variation in those rates has been much reduced as Figure 14.3b shows. The effect of evening out variations has been to *reduce* the strength of the relationship in this case. Often, however, the effect will be to *increase* the value of r. Robinson (1950) gives an example in which $r = 0.203$ with respect to a set of individuals, $r = 0.773$ for aggregation to 48 US states, and $r = 0.946$ for nine major regions. Correlations at any level above that of the smallest possible individual units of analysis are known as **ecological correlations**. Conclusions drawn may be valid at one level of aggregation, but not at another. Notice too the ecological fallacy — a correlation, say, between houses with no inside toilet and per cent New Commonwealth citizens does *not* allow us to conclude that all coloured people live in substandard housing.

Finally, we must consider the special problems that may be posed by percentage data, over and above their tendency to exhibit strong left- or right-skewness, when occurring in the form of a **closed number system**. Suppose that all the land in each of four regions, A, B, C and D, may be classified either as arable, pasture or scrubland. Hypothetical values, in units of 100 km^2 are given in Table 14.4a, as are the correlations between areas of each land type. All correlations are extremely strong, and positive. If areas are now expressed as percentages of the regional totals, the corresponding correlations are strikingly different in two of the three cases (Table 14.4b). Two of the values are now very strongly negative. It has been shown that for a three-variable case involving values that sum to a fixed total (in this case 100%), two of the correlations will always be negative, and one positive, no matter what the correlations between the set of open numbers from which the closed numbers were obtained. Great care is therefore required when interpreting correlations based on the percentages of persons employed in a number of different kinds of industries or falling in a number of different social groups. The problem was originally recognised in geological studies because lithological components in a stratigraphic unit are commonly expressed as percentages of total thickness when carrying out facies studies (Krumbein & Graybill 1965, pp. 242–3).

Table 14.4   Correlation coefficients based on open and closed number systems.

(a) Open number system, land type (100 km²).

Region	Arable (1)	Pasture (2)	Scrub (3)
A	5	2	1
B	7	4	2
C	8	6	4
D	10	8	5

$r_{12} = 0.99, r_{13} = 0.96, r_{23} = 0.89.$

(b) Closed number system, land type (%).

Region	Arable (1)	Pasture (2)	Scrub (3)	Totals
A	63	25	12	100
B	54	31	15	100
C	44	33	23	100
D	43	35	22	100

$r_{12} = -0.97, r_{13} = -0.98, r_{23} = 0.89.$

TESTING THE SIGNIFICANCE OF r

We have assumed so far that any conclusions based on values of r depend upon the judgement and geographic knowledge of the investigator. If a *population* of measurements is used, the correlation coefficient is the true one for that population. This population parameter is known as ρ (rho) to distinguish from the correlation coefficient, r, based on sample data. In practice, the distinction is rarely made unless a test of significance is being carried out with respect to a sample correlation coefficient, r.

For tests of significance to be carried out it is assumed that pairs of measurements are obtained either in random sampling from a population of phenomena such as houses, regions or slopes, or with respect to phenomena representing a realisation of a natural or socio-economic stochastic process. A further assumption is that the joint or bivariate distribution of the population of measurements on X and Y is **bivariate normal**. Finally, X and Y must be measured on an interval or ratio scale, and the relationship in the population should be *linear*. Examples of bivariate normal distributions are given in Figure 14.4. Notice that the *marginal* distributions of X and Y, i.e. the distributions obtained by pushing all the observations at right angles towards the X or Y axes respectively until they 'pile up' along one of those axes, must themselves be normal under these circumstances (although the reverse is not necessarily true).

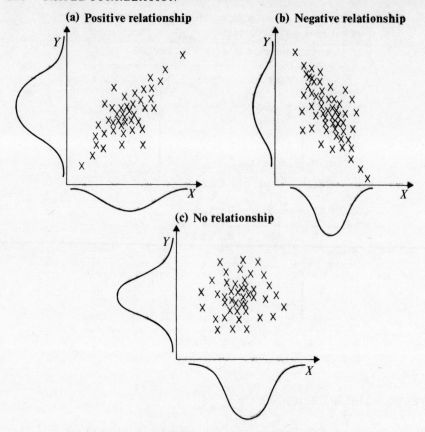

Figure 14.4   Examples of bivariate normal distributions.

We have been shown how to check for normality in Chapter 7, and this distributional assumption places tests of significance for r in the family of parametric techniques.

The commonest test helps us decide whether the correlation coefficient ρ for the population of measurements from which our sample value of r was obtained is likely to be zero or not. If it seems likely that $\rho = 0$, the implications for our study may be considerable. Assuming $\rho = 0$, what is the probability that the observed value of r has been produced purely by chance? The statistical hypotheses are

$$H_0 : \; \rho = 0$$
$$H_1 : \; \rho \neq 0$$

assuming a two-tailed test. For small samples, with $n < 25$, the sample values of r are assumed to follow a t distribution with expected value $\rho = 0$ and estimated standard error

$$s_r = \sqrt{\frac{1 - r^2}{n - 2}}$$

(14.6)

giving the test statistic

$$t_{n-2} = \frac{r - \rho}{s_r} = \frac{r\sqrt{n - 2}}{\sqrt{1 - r^2}}$$

(14.7)

based on $(n - 2)$ degrees of freedom. A sampling experiment carried out by students at Reading, based on $k = 58$ samples, yielded the distribution shown in Figure 14.5.

Bowlby hypothesised *a priori* that the relationship between familiarity and distance, shown in Figure 14.1b would be negative. As earlier calculations show, $r = -0\cdot745$ and $n = 11$.

Set up the statistical hypotheses, and confirm that $t = -3\cdot35$. Use Table A.4 to show that H_0 may be rejected at the 1% level.

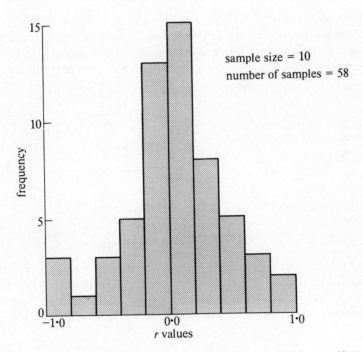

Figure 14.5   Frequency distribution of sample correlation coefficients obtained by experiment from a population for which $\rho \simeq 0\cdot0$.

For large samples, with $n \geqslant 25$, sample values of r follow a normal distribution with expected value $\rho = 0$ and estimated standard error

$$s_r = \frac{1}{\sqrt{n-1}} \qquad (14.8)$$

giving the test statistic

$$z = r\sqrt{n-1} \qquad (14.9)$$

which is tested in the usual way as a standard normal deviate (Table A.3).

Arnett (1971) obtained observations of soil depth (X) and maximum slope angle (Y) by random sampling in an area in Queensland, Australia. The frequency distribution of the Y values indicated a reasonable approximation to normality, but that of the X values showed evidence of left-skewness. As the degree of skewness was mild, values of X were not transformed (see p. 180). However, it may be advisable to employ a stricter significance level, say $\alpha = 0 \cdot 01$.

Arnett found $r = -0 \cdot 68$ with $n = 92$. Show that the z value was $-6 \cdot 49$ and significant at the 1% level.

Sometimes, we may wish to test r against a value of $\rho \neq 0$. As this is rare in geography, we simply state the formulae. Because the sampling distribution of r is not symmetrical about $\rho \neq 0$, r should be transformed according to Fisher's Z transformation (*not* to be confused with the small z of the z test) which is

$$Z = 1 \cdot 1513 \left[\log_{10}(1+r) - \log_{10}(1-r) \right] \qquad (14.10)$$

Z is normally distributed about $\rho \neq 0$ with standard error

$$s_Z = \sqrt{\frac{1}{n-3}} \qquad (14.11)$$

giving the test statistic

$$z = \sqrt{n-3} \times 1 \cdot 1513 \left[\log_{10}(1+r) - \log_{10}(1-r) \right] \qquad (14.12)$$

The test may be employed with any sample size, and values of Z are tabulated in statistical tables such as those by Lindley and Miller (1968) and Neave (1978).

More frequently, we wish to compare r values obtained under two sets of circumstances, or in different regions. The statistical hypotheses are

$$H_0: \rho_1 = \rho_2$$
$$H_1: \rho_1 \neq \rho_2$$

assuming a two-tailed test. Both sample coefficients, r_1 and r_2, must be transformed according to (14.10), and the standard error of $(Z_1 - Z_2)$ is given by

$$s_{Z_1 - Z_2} = \sqrt{\frac{1}{n_1 - 3} + \frac{1}{n_2 - 3}} \qquad (14.13)$$

and

$$z = \frac{Z_1 - Z_2}{\sqrt{\dfrac{1}{n_1 - 3} + \dfrac{1}{n_2 - 3}}} \qquad (14.14)$$

In Bowlby's study, suppose r_1 is the correlation between distance and familiarity for stores in the northern sector, and r_2 that for stores in the southern sector. She found $r_1 = -0.745$, $n_1 = 11$, and $r_2 = -0.616$, $n_2 = 16$. By (14.10) this therefore gives

$$Z_1 = 1.1513 [\log_{10}(0.255) - \log_{10}(1.745)] = -0.961$$
$$Z_2 = 1.1513 [\log_{10}(0.384) - \log_{10}(1.616)] = -0.718$$

Show that $S_{Z_1 - Z_2} = 0.449$, and that the test statistic $z = 0.541$ provides little evidence against H_0.

TRANSFORMATIONS

Transformations of one, or both, variables may be carried out so that the assumptions of *bivariate normality* and *linearity* are more closely approximated. Here we mention only those transformations responsible for normalising observations — any of the transformation discussed in Chapter 12 (pp. 179–80) may be employed. Transformations to overcome the problem of non-linearity are discussed in the next chapter on simple regression analysis (pp. 237–42). Note that it is not necessary for both variables in a correlation analysis to be transformed or, if both are transformed, for them to be transformed in the same way. Transformations should be used with caution, and preferably only if tests of significance are required. It may be rather difficult to decide how to interpret a high correlation between the logarithm of a slope angle and the square root of the thickness of the soil mantle. All conclusions refer to the *transformed* variable(s) and the relationship between them, *not* to the original variables.

CORRELATION AND REGRESSION

These techniques are often discussed together, as they are statistically related. However, they are treated separately here because, first, there is considerable emphasis on estimation as well as upon hypothesis testing in regression; secondly, one of the major statistical models underlying the inferential use of regression differs from the 'bivariate normal' model of correlation analysis; and finally, regression analysis more often assumes a causal relationship between X and Y rather than the various kinds of association largely discussed in this chapter.

EXERCISES

1. Basin area (X) and total stream length (Y) were measured from $1:50\ 000$ maps and $1:30\ 000$ aerial photographs for the ten drainage basins in the Kyogya drainage system in Uganda (Doornkamp and King 1971, p. 25). These values are given below.

Basin area (square miles)	(X)	4·08	4·53	4·92	6·18	6·22	6·54	7·55	8·92	10·16	12·50
Stream length (miles)	(Y)	9·40	10·85	10·69	10·97	11·94	18·69	17·21	18·14	20·14	20·68

Calculate the value of r_S with respect to these observations. How far does it differ from the value obtained for the product—moment correlation coefficient $r = 0·929$? If a test of significance were to be carried out on the value of r_S, how should it be interpreted?

2. Using the values given in Table 14.2, ascertain whether there is a significant correlation between the ranks of items recalled by passengers after the expressway trip and those recalled by drivers.

3. In a study of attitudes of randomly selected finalists at Reading University towards working in each of the 11 standard regions of the United Kingdom, Silvester et al (1978) found that average ranks allotted by students from the South, the Midlands and the North of England were as shown below.

	South	Midlands	North
Northern Ireland	9·5	10	10
Scotland	7	10	5·5
North	9·5	10	1
Yorkshire and Humberside	9·5	8	4
North-west	9·5	6	7
Wales	4	3	10
West Midlands	5·5	4	5·5
East Midlands	5·5	2	8
East Anglia	3	6	10
South-west	2	6	2
South-east	1	1	3

How much evidence is there of agreement between the three sets of rankings?

4. Carey and Mapes (1972) carried out a study of local friendship patterns on newly completed private housing estates in the North Midlands, the variable

of interest, X_1, being the amount of 'within-estate visiting' carried out by housewives. Variables correlated with X_1 were X_2, age of housewife (respondent); X_3, length of residence on the estate; X_4, distance of the last move (i.e. distance in miles from the respondent's previous home); X_5, number of cars in the household; X_6, membership of local social clubs (i.e. number of clubs on the estate to which respondent belongs).

The correlations between X_1 and each of the individual variables were as shown below:

	X_2	X_3	X_4	X_5	X_6
X_1	−0·2424	0·0900	−0·0827	0·2163	0·0113

The number of housewives interviewed was $n = 158$.

(a) Which of the associations is statistically significant? At which level(s) of significance?
(b) Do the directions or signs of the associations appear to correspond, in every case, with those that might be expected on the grounds of commonsense?

5. Carson (1969) obtained $n = 46$ sample observations from different rock types in areas of northern Exmoor and the southern Pennines. The chief variable of interest was the angle of straight slope (X_1), and correlations obtained with other variables were as shown.

	X_2	X_3	X_4	X_5
	Mean rock fragment fraction	Mean silt-clay content of soil	Stream gradient at foot	Mean thickness of waste mantle
X_1 Slope angle	0·67	−0·36	0·28	−0·87

(a) Are all the correlations significant at the 5% level? Justify use of one- or two-tailed tests.
(b) Comment on the signs of the coefficients.
(c) Are the relationships causal, spurious or do they represent mutual interaction of forces?

6. A geographical study was carried out in the early 1960s to investigate the notion that pairs of large cities are spaced farther apart than pairs of smaller cities, all other things being equal (Thomas 1962). A random sample of 89 cities was drawn, and the population size of each of them determined for each census from 1900 up to and including 1950.

The distance to the nearest city of the same population size was measured for each of the 89 cities, and this process repeated for each year of the analysis, i.e. 1900, 1910, 1920, 1930, 1940 and 1950. Observed correlations between distance from a sample city to its 'nearest neighbour' of the same population size (X_1), and population size of the sample city (X_2) are tabulated below.

Year	1900	1910	1920	1930	1940	1950
Correlation Coefficients	0·30	0·64	0·57	0·57	0·65	0·59

(a) It looks as if the correlation between size and distance has remained fairly stable over time. What statistical tests would you carry out in order to provide further evidence? Justify the form of test used. What must you take on trust in order to do the testing?

(b) Carry out the tests. Think carefully about the comparisons that should be made, and devise an efficient way of testing − there is no need to compare every pair of correlation coefficients.

(c) Comment briefly on the results of the tests, and on the sign of the correlations. Could any interesting changes have occurred which are *not* reflected in the results?

REFERENCES AND FURTHER READING

Examples of the use of the ϕ coefficient are given in:

Doornkamp, J. C. and C. A. M. King 1971. *Numerical analysis in geomorphology*. London: Edward Arnold.

Norcliffe, G. B. 1977. *Inferential statistics for geographers*. London: Hutchinson. (pp. 113−15).

For examples of Spearman's rank correlation in human geography, see:

Bowlby, S. R. and J. A. Silk 1978. *Learning by doing: project work in behavioural and social geography*. Reading: University of Reading, Dept of Geography, Geographical Papers.

Carr, S. and D. Schissler 1969. The city as a trip: perceptual selection and memory in the view from the road. *Environ. Behav.* 1(1), 7−35.

Kirby, A. M. and P. J. Taylor 1976. A geographical analysis of voting patterns in the EEC Referendum, 5 June 1975. *Reg. Stud.* 10, 183−91.

In physical geography, see Doornkamp and King (1971, pp. 51−5, 215, 281−4) (full reference above) and:

Fermor, J. 1972. The dry valleys of Barbados: a critical review of their pattern and origin. *Trans Inst. Br. Geogs* 57, 153−65.

There have been numerous uses of the product−moment correlation coefficient, *r*. In human geography, good examples are:

Carey, L. and R. Mapes 1972. *The sociology of planning*. London: Batsford.

Carter, H., H. A. Stafford and M. M. Gilbert 1970. Functions of Welsh towns: implications for central place notions. *Econ. Geog.* 46, 25−38.

Smith, G. C. 1976. The spatial information fields of urban consumers. *Trans Inst. Br. Geogs* (New Series) 1(2), 175−89.

Thomas, E. N. 1962. Stability of distance−population size relationships for Iowa towns from 1900 to 1950. *Lund Stud. Geog.* B, 24, 13−20.

In physical geography, see Doornkamp and King (1971), as above, and:

Carson, M. A. 1969. Models of hillslope development under mass failure. *Geog. Anal.* 1(1), 76−100.

Arnett, R. R. 1971. Slope form and geomorphological process: an Australian example. In *Slopes: form and process*, D. Brunsden (ed.), 81−92. Institute of British Geographers, Special Publication no. 3. (See also the papers by

Furley (pp. 141–64) and Furley and Whitfield (pp. 165–75) in the same volume.)

Transformations are discussed in Carey and Mapes (1972, Appendix III) (see reference above), and closed number systems in:

Krumbein, W. C. and F. A. Graybill 1965. *An introduction to statistical models in geology*. New York: McGraw-Hill. (pp. 51–2, 242–3).

The issues of ecological correlation and modifiable units are treated in Norcliffe (1977, Ch. 11), as referenced above, and:

Blalock, H. M. (Jr) 1964. *Causal inferences in non-experimental research*. Chapel Hill: University of North Carolina Press. (Ch. 4).

Curry, L. 1966. A note on spatial association. *Prof. Geog.* **18**, 97–9.

Robinson, W. S. 1950. Ecological correlation and the behaviour of individuals. *Am. Sociol. Rev.* **15**, 351–7.

Taylor, P. J. 1977. *Quantitative methods in geography: an introduction to spatial analysis*. Boston: Houghton Mifflin.

Examples are taken from:

Meadows, T. 1977. *Relationships between road accident rates and environmental parameters*. Reading: University of Reading, Department of Geography (mimeographed).

Silvester, C., A. Utley and M. Wills 1978. Residential preferences of finalists at Reading University, 1977. In *Learning by doing: projects in behavioural and social geography*, S. R. Bowlby and J. A. Silk (eds). Reading: University of Reading, Department of Geography.

Transformations of the correlation coefficient are given in:

Lindley, D. V. and J. C. P. Miller 1968. *Cambridge elementary statistical tables*. Cambridge: Cambridge University Press.

Neave, H. R. 1978. *Statistics tables*. London: George Allen & Unwin.

Reference is also made to:

Schumm, S. A. 1956. Evolution of drainage systems and slopes in badlands at Perth Amboy, New Jersey. *Bull Geol. Soc. Am.* **67**(5), 597–646.

15 Simple Regression Analysis

INTRODUCTION

In regression analysis the investigator is concerned primarily with establishing the precise mathematical form of the relationship between two variables by fitting a mathematical function to a set of data. It is assumed that both variables are measured on an interval or ratio scale. The first section of this chapter emphasises the role of regression analysis in prediction and forecasting, and shows how estimates may be obtained and 'goodness of fit' ascertained. In the second section, the technique is examined within the inferential framework of the **regression model**. We show how confidence intervals may be placed about estimates and describe tests of the significance of the regression relationship. After this, we describe procedures for dealing with non-linear relationships. Finally, methods for analysing residuals about regression are discussed.

REGRESSION ANALYSIS IN THE CONTEXT OF PREDICTION AND FORECASTING

If we can establish the mathematical form of the relationship between two variables, then values of one variable may be estimated, given those of the other. The simplest relationship that can be specified is a linear relationship as represented by the straight line in Figure 15.1a. Suppose that a planner is interested in the relationship between number of households and the daily number of car trips generated with respect to a set of geographic zones in a city. Such a relationship is potentially valuable if it can be expressed in a quantitative manner, because knowledge of the number of households in a zone at some future date may enable prediction of the number of trips generated from that zone. The variable whose values are to be estimated or predicted is known as the *dependent* variable, Y. In Figure 15.1a, Y represents the dependent variable 'number of daily car trips'. The variable whose values are used to obtain the estimates or predictions is known as the *independent* variable, X, and represents 'number of households' in Figure 15.1a. The graph of this relationship can be used to read off the number of daily car trips corresponding to a given number of households, e.g. a zone of 1000 households generates 800 trips.

It is easier and more accurate to obtain values of Y from an algebraic equation which represents a *straight line*. Such equations are of the form

$$Y = a + bX \tag{15.1}$$

(a) Positive relationship

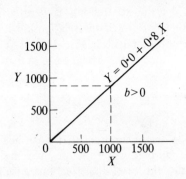

(b) Negative relationship

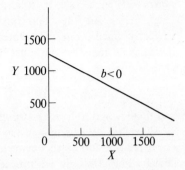

(c) No relationship

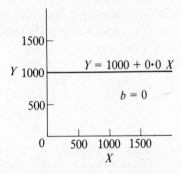

Figure 15.1 Examples of linear relationships between X and Y.

where a is the **intercept term** and represents the point where the straight line cuts the Y axis, and b is the gradient or **slope term**. The slope term controls the rate at which the line rises (or falls) as X increases. The straight line in Figure 15.1a may be represented by the equation

$$Y = 0 + 0 \cdot 8X$$

In this case, $a = 0$ and $b = 0 \cdot 8$. Substituting the value $X = 1000$ in the equation, we obtain

$$Y = 0 + (0 \cdot 8 \times 1000) = 800$$

confirming the result found by the graphical method.

> Population forecasts estimate that the number of households in a particular zone is likely to increase from 800 to 1400. By how much should the number of daily car trips increase?

If the slope term, b, is positive ($b > 0$), the relationship between X and Y is *positive* (Fig. 15.1a); if b is negative ($b < 0$), the relationship is *negative* (Fig. 15.1b). An important special case is that for which the relationship is perfectly linear, but represented by a horizontal straight line so that as X increases, Y remains constant (Fig. 15.1c). The slope in this case is zero, so that

$$Y = a + (0 \times X) = a$$

and there is said to be *no relationship* between X and Y.

Thus far, the discussion has been based on the assumption that a perfect linear relationship exists between X and Y, i.e. all observations lie on the same straight line. How do we proceed if this is not the case? Imperfections arise because it is virtually impossible, even under laboratory conditions, to eliminate entirely the influences exerted by factors other than the independent variable X upon the values of Y. In other words, all other things are not equal, but the investigator usually has reason to believe that if some kind of mathematical function is fitted to the observations, then it will closely approximate the 'true' nature of the relationship between X and Y.

Consider a set of observations taken from a transport study carried out for the City of Belfast (Travers Morgan 1968), plotted in Figure 15.2. The city was divided into 138 zones for the study, so that only 21 zones forming a sector extending westward from the city centre are included here. X represents the

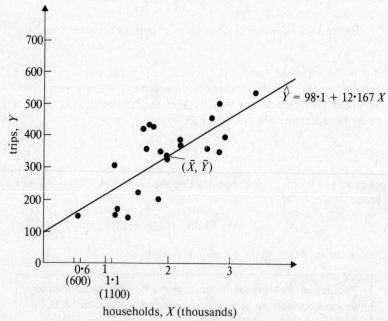

Figure 15.2   Least-squares regression line fitted to Belfast data.

number of households in each zone, and Y the number of car trips leaving the zone each day. The relationship between these variables is clearly imperfect. The **method of least squares** is one procedure which provides a best-fit line, this being a line which, in some sense, lies as close as possible to all the points. Placement of the least-squares line is based on the criterion that the sum of the squared deviations of the points about the line, measured parallel to the Y axis, is minimised. The equation of the least-squares line, known as the **regression equation**, may be written as

$$\hat{Y}_i = a + bX_i \tag{15.2}$$

where the intercept and slope terms a and b are now **regression coefficients** estimated by the least-squares procedure. $\hat{Y}_i$ (Y_i hat) is the value of Y estimated by the regression equation, corresponding to the observed value of X, namely X_i. For the regression coefficients we have

$$b = \frac{\Sigma X_i Y_i - [(\Sigma X_i)(\Sigma Y_i)]/n}{\Sigma X_i^2 - (\Sigma X_i)^2/n} \tag{15.3}$$

$$a = \bar{Y} - b\bar{X} \tag{15.4}$$

Using these formulae, and the quantities given in Table 15.1, we can fit a least-squares line to the Belfast data. (The number of households (X) has been

Table 15.1   Data for the regression of trips (Y) on households (X).

Zone	Y_i	X_i	Y_i^2	X_i^2	$X_i Y_i$	$\hat{Y}_i$	$\hat{Y}_i^2$	$Y_i - \hat{Y}_i$
1	426	17	181 476	289	7242	304·9	92 964	121·1
2	357	26	127 449	676	9282	414·4	171 727	−57·4
3	357	28	127 449	784	9996	438·8	192 545	−81·8
4	342	19	116 964	361	6498	329·3	108 438	12·7
5	534	34	285 156	1156	18 156	511·8	261 939	22·2
6	309	11	95 481	121	3399	231·9	53 778	77·1
7	387	22	149 769	484	8514	365·8	133 810	21·2
8	358	16	128 164	256	5728	292·8	85 732	65·2
9	331	20	109 561	400	6620	341·4	116 554	−10·4
10	388	29	150 544	841	11 252	450·9	203 311	−62·9
11	425	17	180 625	289	7225	304·9	92 964	120·1
12	150	6	22 500	36	900	171·1	29 275	−21·1
13	456	27	207 936	729	12 312	426·6	181 988	29·4
14	374	22	139 876	484	8228	365·8	133 910	8·2
15	489	28	239 121	784	13 692	438·8	192 545	50·2
16	205	18	42 025	324	3690	317·1	100 552	−112·1
17	159	11	25 281	121	1749	231·9	53 778	72·9
18	220	15	48 400	225	3300	280·6	78 736	−60·6
19	424	16	179 776	256	6784	292·8	85 732	131·2
20	168	11	28 224	121	1848	231·9	53 778	−63·9
21	140	13	19 600	169	1820	256·3	65 690	−116·3
Σ	6999	406	2605 377	8906	148 235	6999·3	2489 646	−000·5

$\bar{Y} = 333\cdot29$, $\bar{X} = 19\cdot33$.

expressed in hundreds to simplify the calculations — there are 1700 households in zone 1, 2600 in zone 2 etc.) The slope term is

$$b = \frac{148235 - [(406) \times (6999)]/21}{8906 - (406)^2/21} = \frac{148235 - 135314}{8906 - 7849} = 12 \cdot 224$$

> Show that the intercept term is $a = 97 \cdot 0$ and write out the regression equation.

One of the properties of a least-squares regression line is that it passes through the **centroid** or **bivariate mean** — the point $(\overline{X}, \overline{Y})$ — of the set of observations. As shown in Figure 15.2, $a = 97 \cdot 0$ and the bivariate mean is $(19 \cdot 33, 333 \cdot 29)$.

The procedure outlined above is designed to 'reveal' any linear trend in a set of empirical observations, and may be referred to as 'fitting a least-squares regression line' or 'regressing Y on X'.

Estimation

The regression equation may be used to estimate values of Y by substituting the desired value of X, just as the perfect mathematical relationship was used earlier. The number of trips generated from a zone containing 1000 (10 x 100) households should be

$$\hat{Y} = 97 \cdot 0 + (12 \cdot 224 \times 10) \simeq 219$$

Figure 15.2 and Table 15.1 show that none of the zones used when estimating the regression equation actually contained 1000 households. However, it seems a reasonable assumption that, if a zone containing 1000 households *did* exist, then the number of trips generated from that zone would be approximately 220. For a zone containing any number of households *within* the range of values of X used to obtain the regression equation, it is usually legitimate to **interpolate** values of Y in this way. However, we must assume that the relationship between X and Y remains 'well behaved' in a region between two values of X (such as 600 and 1100 in Figure 15.2) for which, strictly speaking, there is no information. Beyond the range of the data used to obtain the equation, we must be far more cautious. If there are no households resident in a particular zone, the equation predicts that approximately 98 car trips will be generated. This is not an unusual finding in regression studies and it is probably explained here by the fact that attempts to distinguish between households living within a zone, and those not, when interviewing at cordon lines, were not successful. It is dangerous to **extrapolate** beyond the range of observed values of X, because the relationship may become curvilinear or show a 'break of slope' as in Figure 15.3.

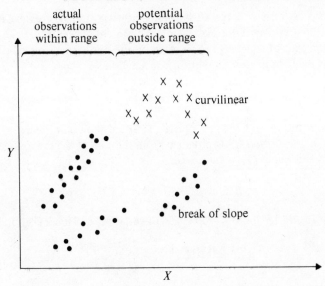

Figure 15.3 Examples of difficulties of extrapolation.

Use the regression equation to estimate values of Y corresponding to $X = 500$ and $X = 5000$. How much confidence have you in these estimates?

Unless the linear relationship between two variables is perfect, the least-squares line produced by the regression of Y on X is *not* the same as that for the regression of X on Y. If there is no linear relationship, the lines cross at right angles; if there is a perfect linear relationship, they coincide. The investigator must clearly specify which variable is dependent, and which independent, before doing the calculations. Difficulties on this score have sometimes provoked heated debates (Adams 1974, Hanlon 1975, Tanner 1974).

Goodness-of-fit and variation explained

The least-squares procedure gives the *best* possible fit of a straight line to the observations, but 'best fit' in this least-squares sense does not necessarily imply a 'good fit'. An index of the goodness-of-fit of the regression line — or of the degree of perfection of the relationship — known as the **coefficient of determination** or R^2 (*R* squared) may be calculated. This coefficient ranges from 0·0, indicating 'no fit' or no relationship, to 1·0, indicating a perfect fit or a perfect relationship. R^2 also represents the proportion of variation in Y 'explained' by

X. The word 'explained' is placed in quotes because statistical explanation and explanation in everyday terms may be rather different concepts.

The *Total* sum of squares (TSS) is given by the sum of the squared deviations of the individual observations about their mean $\bar{Y}$, i.e.

$$\text{TSS} = \Sigma(Y_i - \bar{Y})^2 \tag{15.5}$$

and the elements of this sum of squares (SS) are shown in Figure 15.4a. The 'explained' or *Regression* SS is obtained from the sum of the squared deviations of the predicted values ($\hat{Y}_i$) about $\bar{Y}$, i.e.

$$\text{Regression SS} = \Sigma(\hat{Y}_i - \bar{Y})^2 \tag{15.6}$$

as shown in Figure 15.4b. The *proportion* of variation explained is given by

$$R^2 = \frac{\text{Regression SS}}{\text{Total SS}} = \frac{\Sigma(\hat{Y}_i - \bar{Y})^2}{\Sigma(Y_i - \bar{Y})^2} \tag{15.7}$$

(a) Total sum of squares   **(b) Regression sum of squares**

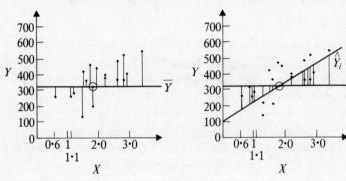

(c) Residual sum of squares

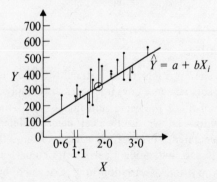

Figure 15.4   Components of sums of squares.

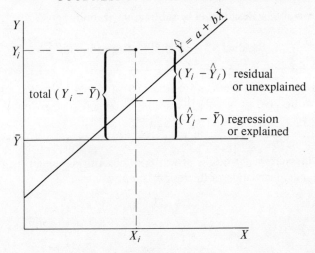

Figure 15.5   Splitting the total sum of squares into regression (explained) and residual (unexplained) components.

The difference between Total SS and Regression SS is 'unexplained' by the regression line and called *Residual* SS. For a typical observation (X_i, Y_i), contributions to Total SS, Regression SS and Residual SS are shown in Figure 15.5. The difference between the observed value Y_i and its corresponding estimate $\hat{Y}_i$ is known as a **residual**, hence

$$e_i = Y_i - \hat{Y}_i \qquad\qquad (15.8)$$

where e_i is the residual value associated with (X_i, Y_i). The sum of the squared residuals may be written

$$\text{Residual SS} = \Sigma e_i^2 = \Sigma(Y_i - \hat{Y}_i)^2$$

individual components of this source of variation being shown in Figure 15.4c.
   To calculate R^2 use

$$\text{Total SS} \quad = \Sigma Y_i^2 - (\Sigma Y_i)^2/n \qquad\qquad (15.9)$$

$$\text{Regression SS} = b\{\Sigma X_i Y_i - [(\Sigma X_i)(\Sigma Y_i)]/n\} \qquad\qquad (15.10)$$

where b is the slope term of the regression equation, and the expression within the curly brackets { } is the numerator of Equation 15.3. Substituting from Table 15.1, we obtain

$$\text{Total SS} \quad = 2605\,377 - (6999)^2/21 = 272\,710\cdot4$$

$$\text{Regression SS} = 12\cdot224 \times 12\,921 = 157\,946\cdot3$$

Notice that we obtain Residual SS by subtraction according to

$$\text{Residual SS} = \text{Total SS} - \text{Regression SS} \qquad \text{15.11)}$$

> Show that $R^2 = 0.58$, and Residual SS $= 114\ 764.1$. Would you describe the fit achieved by the regression line as 'poor', 'moderate' or 'good'?

A direct measure of the degree of scatter of observations about the regression line is the **standard deviation** of the residuals, σ_R, where

$$\sigma_R = \sqrt{\frac{\text{Residual SS}}{n}} \qquad (15.12)$$

and n is the number of pairs of observations. We may also calculate the **coefficient of variation**

$$CV_R = \sigma_R / \overline{Y} \qquad (15.13)$$

so that the degree of scatter is corrected for the magnitude of the quantity being estimated.

> Show that $\sigma_R = 73.93$ and $CV_R = 0.222$ (22.2%) for the Belfast data.

The indices R^2, σ_R and CV_R should only be used as descriptive statistics or summary measures, and have no inferential status as they stand. Slightly different formulae for σ_R and CV_R must be used if statistical inferences are to be made.

CHOICE AND JUDGEMENT IN REGRESSION ANALYSIS

In a prediction or forecasting context, the choice of the dependent variable is usually limited to the one whose behaviour we wish to predict. Choice of the most appropriate independent variable may pose greater problems. Ideally, the independent variable should in some sense be regarded as the 'cause' of the dependent variable, e.g. households produce trips, rather than trips households! Secondly, values of the independent variable should be more readily available than those of the dependent variable, both now and in the future. If values of the dependent variable are just as easy to obtain, the entire exercise wastes time and effort. Many forecasting exercises in transport planning rely on the availability of current figures for office floor space, number of employees, population and the like from local authorities, together with routine forecasts of such values for future dates and places (Bruton 1975, Ch. 3). In a study by Stocking and Elwell (1976), values of erosivity (Y), which are expensive and time consuming to establish, were regressed upon detailed rainfalls (X) available from government-financed meteorological stations. An 'erosivity map' was constructed by applying an estimating equation based on 33 recording points to over 400 stations (see Qu. 1, below).

All too often our knowledge of causal relationships between variables may be so poor that we fall back upon the device of regressing the dependent variable upon any variable to which it might reasonably be related, e.g. regressing changes in employment over a selected period (Y) on employment totals providing at the beginning of that period (X) (see Qu. 3, p. 228) or summer rainfall totals (Y) on those of the immediately preceding spring (X). 'Causes' may even be regressed upon 'effects', if the values of the former are very much easier to obtain than those of the latter (see Qu. 2, p. 228).

Whatever the relationship between X and Y in the senses described above, the most important methodological assumption made when applying the results of a regression analysis (or any other technique used for prediction and forecasting) to other times or places is that the forces or processes underlying the relationship do not change basically. Equations developed to predict changes in phenomena involving socio-economic, political and psychological factors may have a particularly limited range of application. For difficulties experienced in forecasting changes in land use and transport activities see Lowry (1965) and Lee (1973). It has also been suggested in human geography and planning that more emphasis be placed on a qualitative or 'judgemental' approach which explicitly examines the socio-economic and political forces at work (Chapman 1976, Hall 1978).

Finally, it should be remembered that just as there can be 'nonsense correlations' (Ch. 14) so there can be 'nonsense regressions'. This is particularly true of time-series data consisting of observations on two variables over the same period.

EXERCISES I

1. The regression equation obtained by Stocking and Elwell in their study of rainfall and erosivity was

$$\hat{Y}_i = 311 \cdot 38 + 13 \cdot 23 X_i$$

$$R^2 = 0 \cdot 617 \qquad \sigma_R = 1701 \qquad \bar{Y} = 9001$$

where X is mean rainfall annual, Y is annual erosivity. This equation was based on 68 observations. (Note that the number of observations exceeds the number of stations (33) because records were split so that a station with a 10-year record provided two observations each based on a five-year period, a station with a 15-year record three observations, and so on.) Stations were also classified according to their location in one of four regions. It was thought that a separate estimating equation should be used for the stations located in the Eastern Districts, since all the observations for this region fall below the 'all points' line.

(a) Use the observations below, for the Eastern Districts, to regress erosivity (Y) upon rainfall (X). ($n = 11$.)

Erosivity (Y_i)	5·7	8·2	9·4	9·6	9·8	10·2	11·9	11·7	13·4	13·2	13·4
Rainfall (X_i)	7·1	7·2	7·3	7·7	7·6	8·7	9·1	9·5	10·3	10·5	11·2

Values of Y are expressed in thousands of units ($J\ mm^{-1}\ m^{-2}\ h^{-1}$) and values of X in hundreds of units (mm). To obtain an estimating equation in the correct units, remember to multiply the values of a and b by one thousand.

(b) Plot the observations and then calculate the values of a and b according to the method of least squares. Plot the regression line and ascertain whether it provides a good fit to the data both by visual inspection and by calculating R^2 and σ_R. Is the fit achieved by this line to the Eastern Districts observations better than that achieved by the 'all points' line to all 68 observations?

(c) Careful checking of their article shows that Stocking and Elwell in fact regressed rainfall upon erosivity, and then switched the axes! Does this lead to serious error in the case of the observations for the Eastern Districts; for all the observations?

2. When planning hydro-electric, flood-control or irrigation works, it is important to know the past history of stream flows. In one case a dam was being planned in the mid-1940s at Newgate, British Columbia, on the Kootenai River, but stream records there extended back only to 1931 (Ezekiel & Fox 1959). However, records were available for a longer period further downstream over the American border at Libby in Montana. Could flows at Newgate be estimated from those at Libby? The annual flow volume at Newgate represents the dependent variable (Y) and that at Libby the independent variable (X). The observations at the two locations over the period 1931–43 are given below.

	1931	32	33	34	35	36	37	38	39	40	41	42	1943
Newgate (Y_i)	19·7	18·0	26·1	44·9	26·1	19·9	15·7	27·6	24·9	23·4	23·1	31·3	23·8
Libby (X_i)	27·1	20·9	33·4	77·6	37·0	76·6	17·6	35·1	32·6	26·0	27·6	38·7	27·8

$\bar{Y} = 24·96$	$\Sigma Y_i = 324·5$	$\Sigma Y_i^2 = 8742·09$	$\Sigma X_i Y_i = 12931·82$
$\bar{X} = 32·54$	$\Sigma X_i = 478·0$	$\Sigma X_i^2 = 21857·92$	$n = 13$

(a) Plot the observations. Do you think the relationship between X and Y could reasonably be represented by a straight line?

(b) Regress Y on X and also calculate the values of R^2 and σ_R. Does the line provide a good fit?

(c) Annual flow volumes at Libby for the period 1925–30 were 42·0, 24·0, 38·0, 49·4, 24·6, and 24·2 respectively. Estimate the annual volume of flow at Newgate for each of these years. Are you justified in feeling confident in every one of the values estimated?

3. Foot (1974) analysed data on changes in service employment in 23 zones in the Reading sub-region over the period 1961–66. Regressing change in

service employment in each zone (Y) on total service employment in each zone in 1961 (X), he found that

$$\hat{Y}_i = 6 \cdot 167 + 0 \cdot 178 X_i$$

$$R^2 = 0 \cdot 977 \qquad \sigma_R = 165 \cdot 6 \qquad Y = 460 \cdot 1$$

(a) Does the regression line provide a 'good fit' to the observations?
(b) Assuming these results had been available in 1966, do you think the equation should have been used to predict changes in service employment in the Reading area over the period 1966–71? What assumptions should have been made if this were done?

REGRESSION ANALYSIS AND STATISTICAL INFERENCE

We have seen how a perfect linear relationship between two variables can be expressed as a mathematical function and how the least-squares method can be used to fit such functions even when the relationship is imperfect.

The role of statistical inference in regression analysis must now be considered, for the calculated regression equation may pertain to sample realisations of some stochastic process (natural sampling context) or to a random sample of measurements obtained from a statistical population (artificial sampling context). If either is the case, the equation represents an estimate of the true regression equation that would have been obtained from complete knowledge of the process or of the population of measurements. The calculated values of the regression coefficients representing the slope and intercept terms are subject to sampling error, as are the $\hat{Y}_i$, and must be regarded as estimates rather than as parameters. If certain assumptions are fulfilled, confidence intervals may be defined and tests of significance carried out.

The regression model and its underlying assumptions

The **linear regression model** is based upon the assumption that the relationship between X and Y consists of a systematic or deterministic component represented by a straight line, and a random or stochastic component represented by the deviations of the observations about the line. For each observation, the relationship may be written

$$Y_i = \alpha + \beta X_i + \epsilon_i \qquad (15.14)$$

where α and β are parameters representing the true or population intercept and slope terms respectively, and ϵ_i is a stochastic error or **disturbance term** associated with the ith observation.

Two forms of the linear regression model should be distinguished. The '**fixed X' model** implies that the investigator can predetermine the values of X. Consider a study in physical geography where the investigator controls the

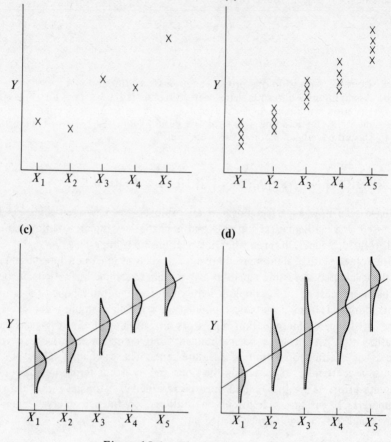

Figure 15.6 The fixed-X model.

discharge rate (X) of water into an experimental flume, and measures some aspect of channel geometry (Y) with respect to each level $X_1, X_2, \ldots$ of X (Fig. 15.6a). Repeating the experiment several times at each value of X, the results might be as shown in Figure 15.6b, due to random variation in the values of Y. A very large number of repeats yield sampling distributions of the values of Y corresponding to each value of X as in Figure 15.6c. This diagram illustrates many of the assumptions underlying the linear regression model.

(1) The relationship between X and Y is linear.
(2) Values of X are fixed and measured without error or, at most, with negligible error.
(3) The disturbance terms ϵ_i are normally distributed with equal variance about the line $Y = \alpha + \beta X$, and each has expected value $E(\epsilon_i) = 0$.

Notice the above assumption implies that the expected value of Y corresponding to any given value of X is $E(Y \mid X_i) = \alpha + \beta X_i$. A further important assumption is now given.

(4) The ϵ_i are statistically uncorrelated. This 'no-correlation assumption' must hold in a number of senses: there should be no trend in the ϵ_i with respect to the true relationship $Y = \alpha + \beta X$ (this is the assumption of no autocorrelation); the ϵ_i should not be correlated with any variable, other than X, which might conceivably be included in the analysis; if the observations are for locations in space there should be no spatial trend or patterning, other than a purely random one, in the ϵ_i (this is the assumption of no spatial autocorrelation).

Figure 15.6c shows a model in which the assumptions are satisfied, although no information is provided on the assumption of spatial autocorrelation.

> Which assumptions of the linear regression model are violated
> by the population of observations in Figure 15.6d?

The ϵ_i may be regarded as the equivalents, in the population of measurements, of the residuals about the regression line mentioned earlier (p. 225) and are interpreted as the result of numerous small influences, including measurement error, which tend to cancel out and thus produce a normally distributed pattern about the line. Our assumption of no correlation (assumption (4)), in its various forms, ensures that all possible sources of *systematic* error or 'disturbance' are eliminated, so that estimates of the intercept and slope parameters α and β are unbiased and follow sampling distributions of minimum variance.

Although the assumptions of the linear regression model may readily be appreciated in terms of the fixed X model, the **'random X' model** is more widely used by geographers. It is assumed either that the values of X and Y represent sets of realisations of a stochastic process in a natural sampling context, or that they pertain to individuals, firms or areas randomly selected from a statistical population. Clearly, the investigator must 'take the values as they come' without determining those of X, a more realistic position in many instances. Otherwise, assumptions are as for the fixed X model. A special case of the random X model is the **bivariate normal model**, requiring the additional assumption of a bivariate normal distribution of the population of (X, Y) values, as was the case with tests of significance in parametric correlation analysis.

Whichever model is employed, assumptions (1) to (4) *must* be fulfilled if standard inferential procedures are to be used.

The investigator should take steps to maximise the chance of compliance with these assumptions. This may be possible through careful choice of experimental procedures or sampling designs. However, many assumptions can only be checked *after* a regression line has been fitted to the data. Full discussion of screening checks and formal tests is given later (pp. 242–8), but it is stressed that a scatter diagram of the observations is one of the simplest and most effective ways to check on many assumptions.

> Check the observations shown in Figure 15.2 and for Question 1, p. 227. Do the sample observations look as though they could have been drawn from a population (or generated by a process) which fulfills the assumptions of the regression model?

Notice there is a tendency either for *all* the assumptions — at least, those relating to linearity and the behaviour of the ϵ_i — to be fulfilled, or none.

The sample regression line

If the observations can be regarded as measurements drawn randomly from a population or generated by a 'natural' stochastic process, then the unknown intercept and slope terms α and β are estimated by $\hat{\alpha}$ and $\hat{\beta}$. The **sample regression line** can be estimated from the sample measurements, and is represented by the equation

$$\hat{Y}_i = \hat{\alpha} + \hat{\beta}X \qquad\qquad (15.15)$$

where $\hat{Y}_i$ is estimated by least squares. In repeated random sampling (holding sample size constant) from a population of (X, Y) values, or in repeated 'runs' of

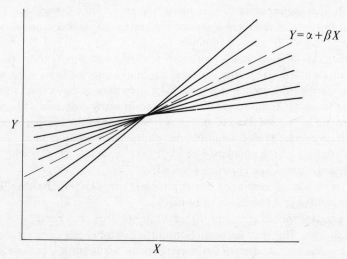

Figure 15.7   Relationship between population and sample regression lines.

a system producing different sets of (X, Y) pairs on each occasion, the frequency distributions of $\hat{\alpha}$ and $\hat{\beta}$ so obtained would be centred on α and β so that $E(\hat{\alpha}) = \alpha$ and $E(\hat{\beta}) = \beta$. Values of $\hat{\alpha}$ and $\hat{\beta}$ derived from any given set of sample observations will therefore differ from those of α and β. It follows that the population line, shown by the broken line in Figure 15.7, and the lines associated with different samples or realisations will differ (full lines in Figure 15.7). Thus, the deviations of the observations about the sample regression line will differ from those of disturbances (ϵ_i) about the population regression line. The deviations about the sample regression line are called **sample residuals**, and denoted by e_i. Thus, the value of an individual observation on Y may be written

$$Y_i = \hat{\alpha} + \hat{\beta}X_i + e_i \qquad (15.16)$$

where $\hat{\alpha}$ and $\hat{\beta}$ provide the least-squares estimates of the population parameters α and β, and e_i is a sample residual value which represents an estimate of the corresponding population disturbance ϵ_i (Fig. 15.17).

Precision of estimates from the sample regression equation

If the equation is to be used for prediction or forecasting, some idea of the accuracy and reliability of estimates is required. The first two measures are based on the sample standard deviation of the residuals, s_R, where

$$s_R = \sqrt{\frac{\text{Residual SS}}{n-2}} \qquad (15.17)$$

and $(n-2)$ is the number of degrees of freedom associated with residual SS (s_R is also known as the **standard error of estimate**). A sample coefficient of variation, cv_R, is given by

$$cv_R = s_R/\overline{Y} \qquad (15.18)$$

Show that $s_R = 77\cdot72$ and $cv_R = 0\cdot233$ for the Belfast data. How do the results compare with those based on σ_R and cv_R (p. 226)?

Confidence intervals may also be constructed about the regression line, and its form with respect to the Belfast data is shown in Figure 15.8 with the 95% limits represented by the broken curves. Interpretation is as established in Chapter 11. Consider the mean or expected value of Y, namely $E(Y_0)$, that corresponds to any given value of X, namely X_0. If samples of n (X, Y) pairs of

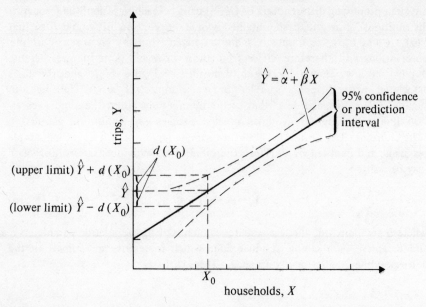

Figure 15.8   Confidence or prediction limits about the regression line for the Belfast data.

measurements were repeatedly taken from a population conforming to the assumptions of the regression model, approximately 95% of the limits constructed about the sample regression lines would contain the true mean value of Y. To construct the limits, an estimate of the standard error of $\hat{Y}_0$ is required. This is given by

$$s(\hat{Y}_0) = s_R \sqrt{\frac{1}{n} + \frac{(X_0 - \bar{X})^2}{\Sigma X_i^2 - (\Sigma X_i)^2/n}} \qquad (15.19)$$

where s_R is the standard error of estimate. The value of $\hat{Y}_0$ follows a t distribution about the population relationship, and so the distance, measured parallel to the Y axis, at which points on each of the curves forming the confidence limits are placed with respect to the sample regression line is given by

$$d(X_0) = t_{n-2,\alpha/2} \times s(\hat{Y}_0) \qquad (15.20)$$

where $(1 - \alpha) \times 100\%$ is the confidence level associated with the limits to be determined, and $t_{n-2,\alpha/2}$ the two-tailed t value, with $(n-2)$ df, associated with the $\alpha \times 100\%$ significance level. The appropriate value for the Belfast data is $t_{19,0\cdot025} = 2\cdot093$ (Table A.4), and limits about the line at the point X_0 are given by

$$\hat{\alpha} + \hat{\beta}X_0 \pm d(X_0) \qquad (15.21)$$

Inspection of Formula 15.19 shows that the second term under the square root vanishes when $X_0 = \bar{X}$, so

$$d(X_0) = d(\bar{X}) = t_{19,0.025} \times s_R/\sqrt{n}$$
$$= 2.093 \times 77.72/\sqrt{21} = 35.50$$

From (15.21), this gives limits of

upper: $97.0 + (12.224 \times 19.33) + 35.50 = 367.69$

lower: $97.0 + (12.224 \times 19.33) - 35.50 = 296.69$

as shown in Figure 15.8.

For $X_0 = 30$, show that the upper 95% limit is 527·84 and the lower limit 399·60.

The form of the limits shows that the error in estimation is at a minimum where $X_0 = \bar{X}$ and may increase considerably away from this point. This result supports earlier comments on the dangers of extrapolation.

Significance tests with respect to the sample regression equation

The statistical significance of a sample regression equation may be assessed using a t test based on the sample slope term $\hat{\beta}$. A test of significance with respect to the sample intercept term $\hat{\alpha}$ may also be carried out.

Although the slope term, $\hat{\beta}$, of the sample regression equation may be non-zero, it is possible that such a value has been generated by chance from a population of (X, Y) pairs of observations for which $\beta = 0$, in which case there is no relationship between X and Y (Fig. 15.1c). The statistical hypotheses to be tested are therefore

$$H_0 : \beta = 0$$
$$H_1 : \beta \neq 0$$

in the case of a two-tailed test. A one-tailed or directional test should be employed if strong *a priori* evidence suggests it. The testing of the estimated slope term $\hat{\beta}$ against $\beta = 0$ is known as testing the overall significance of the regression equation.

Under H_0, repeated sampling yields a distribution of $\hat{\beta}$ which follows a t distribution about an expected value of $\beta = 0$. The test statistic is

$$t_{n-2} = \frac{\hat{\beta} - \beta}{s_{\hat{\beta}}} \tag{15.22}$$

where $(n - 2)$ is the number of degrees of freedom, and $s_{\hat{\beta}}$ the estimated standard error of the sampling distribution of $\hat{\beta}$.

Write out the form of the test statistic under the assumption that $\beta = 0$.

The estimated standard error is given by

$$s_{\hat{\beta}} = \sqrt{\frac{s_R^2}{\Sigma X_i^2 - (\Sigma X_i)^2/n}} \qquad (15.23)$$

where

$$s_R^2 = \frac{\text{Residual SS}}{n - 2} \qquad (15.24)$$

For the Belfast data, show that $s_{\hat{\beta}} = 2\cdot391$ and $t = 5\cdot11$. Does it seem reasonable to reject H_0 (Table A.4)? At what level of significance?

There are also occasions on which it is desirable to test $\hat{\beta}$ against a non-zero value of β (see Qu. 3, p. 249).

The test on the sample intercept term $\hat{\alpha}$ is less widely used, but may be compared profitably with a value of theoretical interest, as in Question 3, p. 249. Once again, a t test is employed, the statistical hypotheses being

$$H_0: \alpha = \alpha^*$$

$$H_1: \alpha \neq \alpha^*$$

assuming a two-tailed test, where α^* is the value of theoretical interest. Under H_0, the sampling distribution of $\hat{\alpha}$ follows a t distribution about an expected value of α^* with a standard error estimated by

$$s_{\hat{\alpha}} = s_R \sqrt{\frac{\Sigma X_i^2}{n\Sigma X_i^2 - (\Sigma X_i)^2}} \qquad (15.25)$$

The test statistic is

$$t = \frac{\hat{\alpha} - \alpha^*}{s_{\hat{\alpha}}} \qquad (15.26)$$

based on $(n - 2)$ degrees of freedom.

Using the data of Table 15.1, show that $s_{\hat{\alpha}}$ = 49·24. Is the
sample intercept term $\hat{\alpha}$ = 97·0 for the relationship between
trips and households significantly different from α^* = 0?

NON-LINEAR RELATIONSHIPS

So far, we have only considered fitting *simple linear* regression models of the
form

$$Y_i = \alpha + \beta X_i + \epsilon_i \tag{15.27}$$

on the assumption that there is a *simple linear* relationship which is linear in
the variables X and Y as well as in the unknown parameters α and β. Relationships such as

$$Y_i = \alpha + \beta X_i^2 + \epsilon_i \tag{15.28}$$

$$Y_i = \alpha + \beta \frac{1}{X_i} + \epsilon_i \tag{15.29}$$

$$Y_i = \alpha + \beta X_i^{-1\cdot7} + \epsilon_i \tag{15.30}$$

which are *non-linear* in the variables but *linear* in the parameters are called
intrinsically linear relationships (Fig. 15.9). This can be shown if we set $Z_i = X_i^2$
and rewrite (15.28) as

$$Y_i = \alpha + \beta Z_i + \epsilon_i$$

which is of the form shown in (15.27).

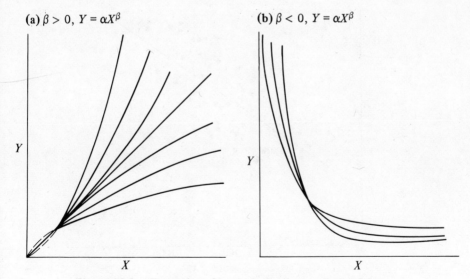

(a) $\beta > 0$, $Y = \alpha X^\beta$ **(b)** $\beta < 0$, $Y = \alpha X^\beta$

Figure 15.9   Examples of intrinsically linear relationships.

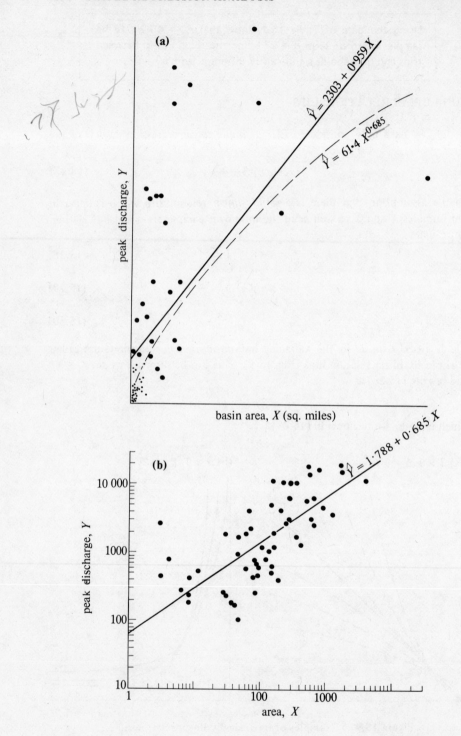

(a)

$\hat{Y} = 2303 + 0.959\,X$

$\hat{Y} = 61.4\,X^{0.685}$

peak discharge, Y

basin area, X (sq. miles)

(b)

$\hat{Y} = 1.788 + 0.685\,X$

peak discharge, Y

area, X

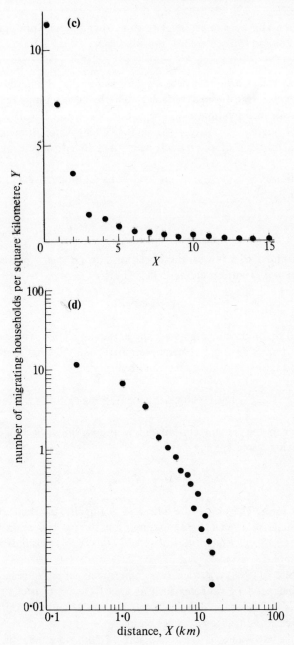

Figure 15.10   Examples of intrinsically linear relationships. (a) Plot of observations of mean annual peak discharge (Y) and basin area (X): untransformed.
(b) Plot of observations shown in Figure 15.10a on 'double-log' paper. (c)
Hägerstrand's study of distance and spatial interaction in Asby: untransformed
observations. (d) Hägerstrand's study of distance and spatial interaction in Asby:
transformed observations.

> Define appropriate variables to show that the relationships in (15.29) and (15.30) are intrinsically linear.

The original values of X have been *transformed* so that Z (and those you have defined) represent **transformed variables**, and the relationships in which they occur are **transformed relationships**. The notion of intrinsic linearity thus widens the scope of regression analysis to embrace many nonlinear relationships between variables. One of the models widely employed by geographers is now considered.

Many relationships in geography are held to be **multiplicative** and therefore of the form (Fig. 15.9)

$$Y = \alpha X^\beta \tag{15.31}$$

Because X is raised to a power, these relationships are called **power functions**. The *regression model* corresponding to (15.31) is

$$Y_i = \alpha X_i^\beta \epsilon_i \tag{15.32}$$

where α and β are unknown parameters and ϵ_i the population error term. Taking logarithms of both sides, and remembering that multiplication involves adding logarithms, and raising to a power their multiplication, we obtain

$$\log Y_i = \log \alpha + \beta \log X_i + \log \epsilon_i \tag{15.33}$$

which is linear in the parameters but not in the variables. Setting $Y_i' = \log Y_i$, $\hat{\alpha}' = \log \alpha$, $X_i' = \log X$ and $\epsilon_i' = \log \epsilon_i$, gives

$$Y_i' = \alpha' + \beta X_i' + \epsilon_i' \tag{15.34}$$

in the desired form. The primes are added to distinguish the transformed terms from their counterparts in (15.32). Assumptions of the regression model now apply to the *transformed* quantities of (15.33) or (15.34) as is illustrated in the following example.

Nash and Shaw (1966) obtained data on mean annual peak discharge (Y) and a variety of independent variables such as area (X) with respect to 57 gauged drainage basins in England and Wales, hoping to derive an equation to estimate flood discharge from ungauged drainage basins.

A plot of the raw values is given in Figure 15.10a. Note the large number of observations near the origin. As shown by the envelope drawn around the points, the relationship seems to be one in which Y increases at a decreasing rate as X increases, and so might well be approximated by a power function of the form $Y = \alpha X^\beta$.

> Is there any evidence of violation of the assumptions under-
> lying the sample linear regression model in Figure 15.10a?

Plotting the original observations on special graph paper, with both axes cali-
brated on a logarithmic scale, effectively transforms the variables and their
relationship. The assumptions of the regression model should be fulfilled by the
transformed observations if our choice of a multiplicative model is correct.

> How well do the observations of Figure 15.10b appear to
> comply with these assumptions?

To fit the multiplicative model, we regress $Y' = \log Y$ on $X' = \log X$. It is
found that $\hat{\alpha}' = 1 \cdot 788$ and $\hat{\beta}' = 0 \cdot 685$ so that

$$\widehat{\log Y_i} = 1 \cdot 788 + 0 \cdot 685 \log X_i$$

and $R^2 = 0 \cdot 556$. This regression line should be plotted with care, bearing in
mind that all values are logarithms, and $\log(1 \cdot 0) = 0 \cdot 0$. To use the sample
regression equation for estimation purposes, we take antilogarithms obtaining

$$\hat{Y} = 61 \cdot 4 X^{0 \cdot 685}$$

and substitute appropriate values of X. Confidence or prediction limits may be
obtained using Equations 15.20 and 15.21, substituting $X_i' = \log X_i$ and $Y_i' =
\log Y_i$ for X_i and Y_i.

The results of a regression analysis based on transformed variables must be
interpreted with care, always bearing in mind the nature of the transformation
used. Nevertheless, the results obtained are frequently valuable, especially when
compared with those derived from an equivalent simple linear model. Regression
of Y upon X produced the estimating equation

$$\hat{Y} = 2302 \cdot 845 + 0 \cdot 959 X$$

and $R^2 = 0 \cdot 342$. When plotted in Figure 15.10a, the untransformed relationship
appears much less satisfactory, avoiding the cluster of observations near the
origin and estimating a mean annual discharge of 2303 for a drainage basin of
zero area. The curvilinear relationship, on the other hand, neatly bisects the
cluster and intercepts the Y axis where $X = 1 \cdot 0$ at the more plausible value of
$61 \cdot 4$.

Human geographers have often used logarithmic transformations of both
variables to estimate the form of **distance decay functions**, a classic example
being given by Hägerstrand's work on spatial diffusion. Characteristics of the

242 SIMPLE REGRESSION ANALYSIS

mean information field for his Asby study were derived from the regression of the logarithm of interaction intensity (X') upon the logarithm of distance (Y'). A plot of the untransformed observations is shown in Figure 15.10c and of the transformed observations in Figure 15.10d.

EXAMINATION OF RESIDUALS

Examination of residuals is a crucial activity in regression analysis. In a *descriptive* context, it may provide useful information on special cases, or reveal trends previously concealed. In an *inferential* context, the sample residual values are inspected to ascertain whether the population disturbances fulfill the assumptions of the regression model, and thereby represent a truly random component about the systematic relationship identified by the regression line. If assumptions are fulfilled, it does not pay us to investigate further, and the unexplained or residual sum of squares is regarded as 'random noise'. The sample estimates $\hat{\alpha}$ and $\hat{\beta}$ may also be regarded as unbiased, i.e. $E(\hat{\alpha}) = \alpha$, $E(\hat{\beta}) = \beta$, and of minimum variance, i.e. the variances of the sampling distributions of $\hat{\alpha}$ and $\hat{\beta}$ are minimised. However, if any evidence of a systematic pattern in the population disturbances is revealed, action must be taken to remove it, otherwise statistical inferences are invalid.

Let us first consider residuals in a descriptive context, calculated from

$$e_i = Y_i - \hat{Y}_i = Y_i - (a + bX_i) \tag{15.34}$$

Residuals associated with observations lying above the regression line are known as **positive** residuals, representing **underestimation** of the observed by the fitted value, and those associated with observations lying below the line as **negative** residuals, representing **overestimation** of the observed by the fitted value. Regressing change in service employment over the period 1961–66 (Y) on service employment in 1961 (X) for 23 zones in the Reading subregion, Foot (1974) obtained the equation

$$\hat{Y}_i = 6{\cdot}167 \times 0{\cdot}178X_i \tag{15.35}$$

Fitted and residual values are given in Table 15.2.

> Use Equations 15.34 and 15.35 to show that the residual values, ignoring sign, are 32, 9 and 643 respectively for zones 1, 8 and 15. State whether the residuals are positive or negative.

Reference to the residual values of Table 15.2 and the map in Figure 15.11, showing zones with positive residuals as black and those with negative residuals

Table 15.2   Service employment regression − observations, fitted values of Y and residuals.

Zone	Change in service employment (1961−66) (Y)	Service employment (1961) (X)	Fitted values ($\hat{Y}_i$)	Residuals ($Y_i - \hat{Y}_i$)
1	5187	29 288	5219	-32
2	401	2167	392	9
3	456	2462	444	11
4	310	1672	304	6
5	816	4409	791	24
6	699	2493	450	248
7	535	2827	509	25
8	141	707	132	9
9	0	200	42	−42
10	21	903	167	−146
11	266	1056	194	72
12	124	797	148	−24
13	112	1009	186	−74
14	48	427	76	−34
15	984	1880	341	643
16	9	234	48	−39
17	0	227	47	−47
18	0	570	108	−108
19	129	834	155	−26
20	6	399	77	−71
21	11	369	72	−61
22	409	3091	556	−148
23	18	550	104	−86

as white, provides useful insights. **Outliers**, or observations lying well above or below the majority of observations, are associated with zones 6 and 15 ($e_6 = +248$, $e_{15} = +623$). These represent an area in South Reading in which there had been much recent expansion of office space, and Wokingham urban district, respectively. Most of the areas with negative residuals are rural, as we might expect.

$$\boxed{\text{Why?}}$$

A good example of the descriptive use of residuals is provided by the regression of channel slope (Y) on bankfull discharge (X) (Leopold *et al.* 1964, p. 293) (Fig. 15.12) where, in general, braided and straight channels appear as positive residuals, indicating steeper slopes than comparable meandering channels which appear as negative residuals.

Let us now consider residuals in an inferential context. A scatter diagram of the observations of X and Y may reveal violations of assumptions, as described earlier. However, a scatter diagram known as a **residual plot** may also be constructed in which the residual values, obtained from Equation 15.34 (but

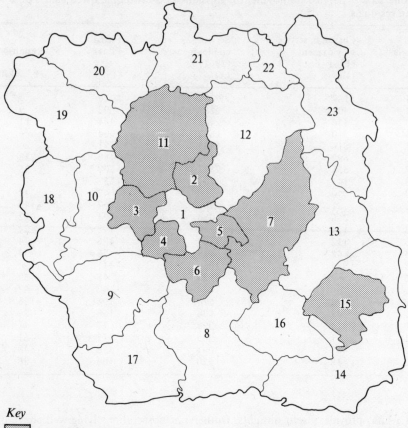

Key

☐ denotes positive residual

Ⓐ zones 1, 2, 3, 4, 5, 6 and 11 are mostly within the built-up area of Reading

Ⓑ zone 15 represents the Wokingham urban area

Ⓒ other zones are predominantly rural

Figure 15.11   Residuals about regression in the Reading sub-region (from Foot 1974, p. 8).

substituting $\hat{\alpha}$ and $\hat{\beta}$ for a and b) are plotted on the vertical axis, and the fitted values of Y, namely the $\hat{Y}_i$, along the horizontal axis. This has been done for the service employment data in Figure 15.13. Residual plots may be difficult to draw, as in this case, because one or two observations have far higher values than any of the others.

Assumptions of normality and equal variance appear to be violated here because the regression line has been unduly influenced by the one extreme observation for zone 1. There is also evidence of autocorrelation among the

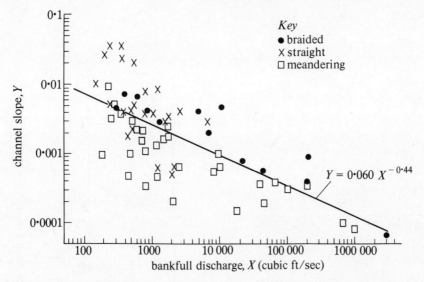

Figure 15.12   Classification of residuals about regression of channel slope (Y) on bankfull discharge (X) (from Leopold, Wolman & Miller 1964, p. 293).

residuals because, starting from the origin, there is a run of negative residuals, followed by a run of positive residuals. We can test for autocorrelation in the residuals using the runs test as previously described in Chapter 6, letting A = positive residual (including zero), and B = negative residual. The statistical hypotheses are

H_0: the population disturbances are not autocorrelated

H_1: the population disturbances are autocorrelated

Given that the sequence of residuals can be coded

BBBBBBBBBBBBBB A A A A A A A B AB

perform a runs test, and confirm that there is sufficient evidence to reject H_0 at the 5% level.

On these grounds, we should be ill advised to use Equation 16.1 for estimation purposes. It is interesting to note that the value of R^2 associated with the equation is 0.977 or 97.7%. A very high proportion of variation explained does *not* necessarily mean we have an equation that fulfills the assumptions of the regression model. Under those circumstances, it is customary *either* to fit a different kind of function, as described in the previous section, *or* to search for another variable to expand the least-squares equation to include two independent variables in a multiple regression analysis. If the latter course is taken, a

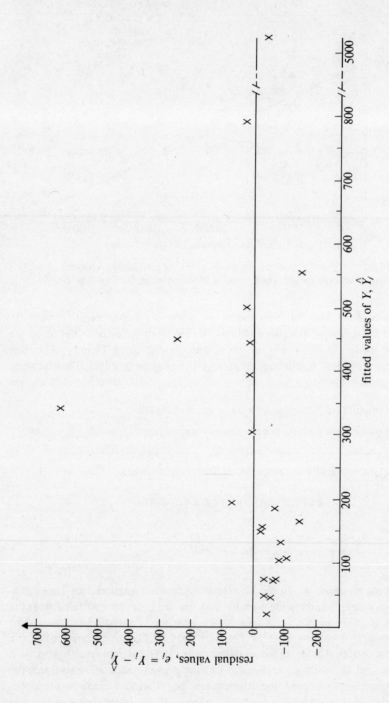

Figure 15.13   Residual plot – change in service employment residuals against fitted values of Y.

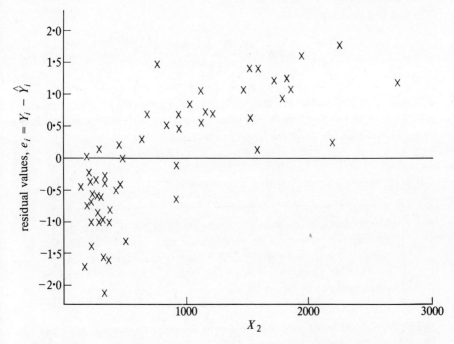

Figure 15.14   Residual plot – log discharge residuals against basin slope (X_2).

second type of plot of use in the analysis of residuals is that of the residual values against those of another independent variable. Figure 15.14 plots residuals from the regression of log(annual peak discharge) (Y) on area (X_1) for the 57 drainage basins studied by Nash and Shaw (1966) against basin slope (X_2). A strong trend is evident, suggesting violation of the assumption of no correlation between the residuals and any other independent variable worthy of inclusion in a multiple regression equation. Because of obvious non-linearity, the logarithm of X_2 should be taken before inclusion in the analysis.

Where the observations are for contiguous areas, or for sample points in space, the residual values may be plotted on a map which is then examined for any visual evidence of a non-random arrangement. Referring once again to the residuals about the service employment equation shown in Figure 15.11, it is clear that change in service employment was overestimated for most of the peripheral and mainly rural areas in the Reading sub-region, suggesting that one variable which in some way measures this characteristic, or another which simply measures accessibility to Reading, may be worth adding to the original equation. Alternatively, the wrong function may have been fitted.

A formal check for *spatial autocorrelation* of residual values may also be carried out. Although by no means the most satisfactory test under these circumstances, the simple contiguity test discussed in Chapter 9 based on the *non-free* sampling assumption may be used. The test statistic recommended by

Cliff and Ord (1973, p. 119) is

$$z(BW) = \frac{J(BW) - \mu(BW)}{\sigma(BW)} \tag{15.36}$$

where areas with positive residuals are coded black (B), those with negative residuals white (W), $z(BW)$ is a standard normal deviate, $J(BW)$ the observed and $\mu(BW)$ the expected numbers of BW joins respectively, and $\sigma(BW)$ the standard error of the number of BW joins. The statistical hypotheses are

H_0: the residuals show no spatial autocorrelation

H_1: the residuals do show spatial autocorrelation

> Check that the assumptions underlying the use of the con-
> tiguity test are fulfilled, and show that $\mu(BW) = 26 \cdot 09$,
> $\sigma(BW) = 3 \cdot 242$, and $z(BW) = 0 \cdot 268$. How do you interpret this
> result after consulting Table A.3? Does it accord with the
> visual evidence of Figure 15.11?

This rather odd result may well be due to the fact that zones 1, 7, 11 and 12 are each connected to almost one third of all the zones in the region. As noted in Chapter 9, the test may be unreliable if a small number of zones dominate the join system. It is perhaps rather appropriate to end this section with a further warning on the dangers of changing spatial patterns into non-spatial indices.

EXERCISES II

1. Bowlby (1972) investigated the relationship between familiarity (Y) and distance (X) with respect to 36 randomly selected supermarkets in and around Evanston, Illinois. She hypothesised that the relationship between familiarity and distance would be negative, and also that the fall-off of knowledge with distance (i.e. the distance–decay function) would be steeper for supermarkets not located between the individuals' homes and Evanston's CBD than it would be for those located in, or in the direction of the CBD. For supermarkets located toward the CBD it was found that

$$Y = 22 \cdot 2 - 4 \cdot 78X \qquad R^2 = 0 \cdot 58 \qquad n = 25$$

For supermarkets located away from the CBD, the observations were as follows.

Familiarity (Y_i)	22	20	15	18	16	16	20	18	18	11	15
Distance (X_i)	0·14	1·25	1·61	1·70	1·77	1·81	1·87	1·97	2·59	2·72	3·18

(a) Fit a simple linear regression model to the data. Is the sample regression line significant?

(b) Is Bowlby's hypothesis with respect to the distance decay functions correct?

2. Use the quantities previously calculated in Question 1, p. 227 to determine whether the sample regression equation relating erosivity (Y) and rainfall (X) is significant.

3. In a student project, Christodoulou and Lewis (1976) asked 40 randomly selected third-year students from the same hall of residence to estimate distances between 20 pairs of points. Observations on the dependent variable Y were calculated as the means of the distance estimates with respect to each pair of points. These values, together with those of X (actual distance) and the other quantities necessary to carry out the analysis are given below.

Y_i (perceived)	X_i (actual)	Y_i (perceived)	X_i (actual)
0·8	0·9	0·9	0·8
2·0	1·3	3·1	2·9
1·2	0·6	2·4	2·2
0·6	0·6	1·1	0·7
0·5	0·5	1·8	1·6
1·0	0·9	1·6	0·8
2·6	1·9	1·2	1·2
1·3	1·2	2·4	2·3
0·7	0·7	1·2	1·0
1·8	1·2	0·9	0·6

$\bar{X} = 1·195$, $\bar{Y} = 1·455$, $\Sigma X_i^2 = 36·93$, $\Sigma Y_i^2 = 52·51$,
$\Sigma X_i = 23·9$, $\Sigma Y_i = 29·1$, $\Sigma X_i Y_i = 43·32$, $n = 20$.

(a) Plot the observations and fit a simple linear regression model to them.
(b) If distances are estimated accurately, then the appropriate regression model is

$$Y_i = 0·0 + 1·0 X_i + \epsilon_i$$

or

$$Y_i = X_i + \epsilon_i$$

Carry out the appropriate tests of significance, one on the slope term, the other on the intercept term, in order to ascertain whether the model stated above is appropriate and the distances estimated accurately.

4. A number of psychological studies (e.g. Ekman and Bratfisch 1965) have shown that the relationship between the degree of emotional involvement (Y) shown by a group of individuals with respect to various cities and the perceived or subjective distance (X) from their home town to those cities may be described by

$$Y = \alpha X^{-0·5}$$

This relationship has been dubbed 'the inverse square-root law' (since $X^{-0·5} = 1/\sqrt{X}$). A further study was carried out by Dornic (1967) in

Czechoslovakia, and the following observations with respect to 10 cities obtained.

	Y_i (emotional involvement)	X_i (subjective distance)
Belgrade	2·82	1·00
Geneva	2·20	2·08
Copenhagen	2·64	1·38
Stockholm	2·21	1·84
Athens	2·21	2·10
Helsinki	2·30	1·45
Alger	1·43	3·64
Delhi	1·38	4·68
Tokyo	1·10	8·30
Buenos Aires	1·00	10·07

The subjective distance scores (X_i) were derived from a psychological scaling technique.

(a) What kind of regression model should be fitted if we wish to test the inverse square-root law with respect to these observations?
(b) Carry out the necessary calculations and fit the model.
(c) Do the results seem to support the law?

5. Langbein (1963) deduced that downstream changes in the width of an 'ideal estuary' should be related to discharge by a function of the form

$$Y = \alpha X^{0.71}$$

where Y represents channel width, and X discharge. Field investigation produced the following observations (Myrick & Leopold 1963). ($n = 12$.)

Y_i (width)	X_i (discharge)
15	20
15	31
17	42
29	91
30	110
42	125
40	200
90	260
80	300
85	330
140	360
140	500

Width is measured in feet, discharge in cubic feet per second.

(a) Write down the form of the regression model that should be fitted to these observations if Langbein's theory is to be tested.
(b) Fit the regression model. Is the sample regression equation significant, and do the results seem to bear out the theory?

REFERENCES AND FURTHER READING

Examples of forecasting applications in transportation and land use planning may be found in:

Bruton, M. J. 1975. *Introduction to transportation planning.* London: Hutchinson.

Chisholm, M., A. E. Frey and P. Haggett (eds) 1971. *Regional forecasting.* London: Butterworth.

Douglas, A. A. and R. J. Lewis 1971. Trip generation techniques: household least-squares regression analysis. *Traffic Eng. Control* 12, 477–9.

Foot, D. H. S. 1974. *A comparison of some land-use allocation/interaction models.* Reading: University of Reading, Dept of Geography, Geographical Papers no. 31.

Difficulties are discussed in:

Lee, D. B. 1973. Requiem for large-scale models. *J. Am. Inst. Plann.* 39, 163–78.

Lowry, I. S. 1965. A short course in model design. *J. Am. Inst. Plann.* 31, 158–65.

For reading on the car ownership forecasting controversy, see:

Adams, J. G. U. 1974. Saturation planning. *Town Cty Plann.* December.

Hanlon, J. 1975. Statisticians' motorway madness. *New Scientist* December 11, 648–9.

Tanner, J. C. 1974. *Forecasts of vehicles and traffic in Great Britain.* Transp. Road Res. Lab. Report no. 650.

Forecasting and estimation problems of interest to physical geographers are considered in:

Nash, J. E. and B. L. Shaw 1966. Flood frequency as a function of catchment characteristics. In *Proc. Symp. on River Flood Hydrology*, 115–36. London: Institution of Civil Engineers.

Newson, M. D. 1975. *Flooding and flood hazard in the United Kingdom.* Oxford: Oxford University Press. (Ch. 2).

Stocking, M. A. and H. A. Elwell 1976. Rainfall erosivity over Rhodesia. *Trans Inst. Br. Geogs* (New Series) 1(2), 231–45.

General treatment of a 'judgemental' approach to forecasting may be found in:

Chapman, G. 1976. Economic forecasting in Britain 1961–1975. *Futures* 8(3), 254–60.

An example is:

Hall, P. (ed.) 1978. *Europe 2000.* London: Duckworth.

For other applications in human geography, in which threshold population sizes are estimated for selected central place functions, see:

Berry, B. J. L. and W. L. Garrison 1958. The functional basis of the central place hierarchy. *Econ. Geog.* 34, 145–54.

Spatial interaction studies are reviewed in:

Olsson, G. 1965. *Distance and human interaction: a review and bibliography.* Philadelphia: Regional Science Research Institute, Bibliographic Series no. 2.

Accessible examples are:

Chisholm, M. and P. O'Sullivan 1973. *Freight flows and spatial aspects of the British economy.* Cambridge: Cambridge University Press.

Starkie, D. N. M. 1967. *Traffic and industry.* London: Weidenfeld and Nicolson/London School of Economics.

Distance decay functions and urban population density gradients are considered in:

Berry, B. J. L., J. W. Simmons and R. J. Tennant 1963. Urban population densities: structure and change. *Geog. Rev.* 53, 389–405.

Warnes, A. M. 1975. Commuting towards city centres: a study of population and employment density in Liverpool and Manchester. *Trans Inst. Br. Geogs* **64**, 77–96.

A useful discussion of the variety of non-linear models that may be fitted to various models of distance decay and spatial interaction is:

Taylor, P. J. 1975. *Distance decay models in spatial interaction*. Concepts and Techniques in Modern Geography (CATMOG), no. 2. Norwich: Geo. Abstracts.

For other applications in physical geography, see papers by Bridges *et al*, Furley *et al* and Statham on slope forms and processes in:

Brunsden, D. (ed.) 1971. *Slopes: form and process*. Inst. Br. Geogs, Special Publication no. 3.

On fluvial processes, see:

Ferguson, R. I. 1973. Channel pattern and sediment type. *Area* 5(1), 38–41.
Gregory, K. J. and D. E. Walling 1973. *Drainage basin form and process – a geomorphological approach*. London: Edward Arnold.

Many other references are cited in:

Chorley, R. J. 1966. The application of statistical methods in geomorphology. In *Essays in geomorphology*, G. H. Dury (ed.), 340–7. London: Heinemann.

References and examples are provided by:

Doornkamp, J. C. and C. A. M. King 1971. *Numerical analysis in geomorphology*. London: Edward Arnold.

The analysis and mapping of residuals is considered in:

Thomas, E. N. 1968. Maps of residuals from regression. In *Spatial analysis*, B. J. L. Berry and D. F. Marble (eds), 326–52. Englewood Cliffs (New Jersey): Prentice-Hall.

For a treatment of spatially autocorrelated residuals, see:

Cliff, A. D. and J. K. Ord 1973. *Spatial autocorrelation*. London: Pion.

Examples may be found in:

Johnston, R. J. 1976. Political behaviour and the residential mosaic. In *Social areas in cities*. Vol. 2: *Spatial perspectives on problems and policies*, D. T. Herbert and R. J. Johnston (eds) 74–86. Chichester: Wiley.
Keeble, D. E. and D. P. Hauser 1972. Spatial analysis of manufacturing growth in outer south-east England 1960–67: 2. Method and results. *Reg. Stud.* 6(1), 11–36.
Taafe, E. J., R. L. Morrill and P. R. Gould 1963. Transport expansion in underdeveloped countries: a comparative analysis. *Geog. Rev.* **53**, 503–29.

A good introduction to multiple regression analysis is given in:

Ferguson, R. I. 1978. *Linear regression in geography*. Concepts and Techniques in Modern Geography (CATMOG), no. 15. Norwich: Geo. Abstracts.

Data sources for examples and exercises, other than those already mentioned, are:

Bowlby, S. R. 1972. *Spatial variation in consumers' information levels*. Evanston: Northwestern University, The Transportation Centre, Household Travel Behaviour Study Report no. 4.
Christodoulou, F. and A. D. Lewis 1976. *Distance perception of Reading University students*. Reading: University of Reading, Dept of Geography, Unpublished Behavioural Geography Project.
Dornic, S. 1967. Subjective distance and emotional involvement: a verification of the exponent invariance. *Rep. Psychol. Labs, Stockholm* **237**, 1–7.
Ekman, G. and O. Bratfisch 1965. Subjective distance and emotional involvement: a psychological mechanism. *Acta Psychol.* **24**, 430–7.

Ezekiel, M. and K. A. Fox 1959. *Methods of correlation and regression analysis.* New York: Wiley.

Langbein, W. B. 1963. A theory for river channel adjustment. *Trans Am. Soc. Civil Eng.* **130**, 679–94.

Leopold, L. B., M. G. Wolman and J. P. Miller 1964. *Fluvial processes in geomorphology.* San Francisco: W. H. Freeman.

Myrick, R. M. and L. B. Leopold 1963. *Hydraulic geometry of a small tidal estuary.* United States Geol. Survey Prof. Paper, no. 422–B.

Travers Morgan, R. 1968. *Travel in Belfast.* Belfast: Belfast Corporation.

APPENDICES

Statistical Tables

Table A.1 Critical values of the chi-square distribution.

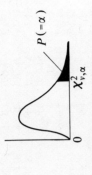

v \ $P(=\alpha)$	0·995	0·990	0·975	0·950	0·900	0·500	0·100	0·050	0·025	0·010	0·005
1	$0·0^4393$	$0·0^3157$	$0·0^3982$	0·00393	0·0158	0·455	2·706	3·841	5·024	6·635	7·879
2	0·0100	0·0201	0·0506	0·103	0·211	1·386	4·605	5·991	7·378	9·210	10·60
3	0·0717	0·115	0·216	0·352	0·584	2·366	6·251	7·815	9·348	11·34	12·84
4	0·207	0·297	0·484	0·711	1·064	3·357	7·779	9·488	11·14	13·28	14·86
5	0·412	0·554	0·831	1·145	1·610	4·351	9·236	11·07	12·83	15·09	16·75
6	0·676	0·872	1·237	1·635	2·204	5·348	10·64	12·59	14·45	16·81	18·55
7	0·989	1·239	1·690	2·167	2·833	6·346	12·02	14·07	16·01	18·48	20·28
8	1·344	1·646	2·180	2·733	3·490	7·344	13·36	15·51	17·53	20·09	21·95
9	1·735	2·088	2·700	3·325	4·168	8·343	14·68	16·92	19·02	21·67	23·59
10	2·156	2·558	3·247	3·940	4·865	9·342	15·99	18·31	20·48	23·21	25·19
11	2·603	3·053	3·816	4·575	5·578	10·34	17·28	19·68	21·92	24·72	26·76
12	3·074	3·571	4·404	5·226	6·304	11·34	18·55	21·03	23·34	26·22	28·30
13	3·565	4·107	5·009	5·892	7·042	12·34	19·81	22·36	24·74	27·69	29·82
14	4·075	4·660	5·629	6·571	7·790	13·34	21·06	23·68	26·12	29·14	31·32
15	4·601	5·229	6·262	7·261	8·547	14·34	22·31	25·00	27·49	30·58	32·80
16	5·142	5·812	6·908	7·962	9·312	15·34	23·54	26·30	28·85	32·00	34·27
17	5·697	6·408	7·564	8·672	10·09	16·34	24·77	27·59	30·19	33·41	35·72
18	6·265	7·015	8·231	9·390	10·86	17·34	25·99	28·87	31·53	34·81	37·16
19	6·844	7·633	8·907	10·12	11·65	18·34	27·20	30·14	32·85	36·19	38·58
20	7·434	8·260	9·591	10·85	12·44	19·34	28·41	31·41	34·17	37·57	40·00

Table A.1 (continued)

v \ $P(=\alpha)$	0·995	0·990	0·975	0·950	0·900	0·500	0·100	0·050	0·025	0·010	0·005
21	8·034	8·897	10·28	11·59	13·24	20·34	29·62	32·67	35·48	38·93	41·40
22	8·643	9·542	10·98	12·34	14·04	21·34	30·81	33·92	36·78	40·29	42·80
23	9·260	10·20	11·69	13·09	14·85	22·34	32·01	35·17	38·08	41·64	44·18
24	9·886	10·86	12·40	13·85	15·66	23·34	33·20	36·42	39·36	42·98	45·56
25	10·52	11·52	13·12	14·61	16·47	24·34	34·38	37·65	40·65	44·31	46·93
26	11·16	12·20	13·84	15·38	17·29	25·34	35·56	38·89	41·92	45·64	48·29
27	11·81	12·88	14·57	16·15	18·11	26·34	36·74	40·11	43·19	46·96	49·64
28	12·46	13·56	15·31	16·93	18·94	27·34	37·92	41·34	44·46	48·28	50·99
29	13·12	14·26	16·05	17·71	19·77	28·34	39·09	42·56	45·72	49·59	52·34
30	13·79	14·95	16·79	18·49	20·60	29·34	40·26	43·77	46·98	50·89	53·67
31	14·46	15·66	17·54	19·28	21·43	30·34	41·42	44·99	48·23	52·19	55·00
32	15·13	16·36	18·29	20·07	22·27	31·34	42·58	46·19	49·48	53·49	56·33
33	15·82	17·07	19·05	20·87	23·11	32·34	43·75	47·40	50·73	54·78	57·65
34	16·50	17·79	19·81	21·66	23·95	33·34	44·90	48·60	51·97	56·06	58·96
35	17·19	18·51	20·57	22·47	24·80	34·34	46·06	49·80	53·20	57·34	60·27
36	17·89	19·23	21·34	23·27	25·64	35·34	47·21	51·00	54·44	58·62	61·58
37	18·59	19·96	22·11	24·07	26·49	36·34	48·36	52·19	55·67	59·89	62·88
38	19·29	20·69	22·88	24·88	27·34	37·34	49·51	53·38	56·90	61·16	64·18
39	20·00	21·43	23·65	25·70	28·20	38·34	50·66	54·57	58·12	62·43	65·48
40	20·71	22·16	24·43	26·51	29·05	39·34	51·81	55·76	59·34	63·69	66·77
45	24·31	25·90	28·37	30·61	33·35	44·34	57·51	61·66	65·41	69·96	73·17
50	27·99	29·71	32·36	34·76	37·69	49·33	63·17	67·50	71·42	76·15	79·49
60	35·53	37·48	40·48	43·19	46·46	59·33	74·40	79·08	83·30	88·38	91·95
70	43·28	45·44	48·76	51·74	55·33	69·33	85·53	90·53	95·02	100·4	104·2
80	51·17	53·54	57·15	60·39	64·28	79·33	96·58	101·9	106·6	112·3	116·3
90	59·20	61·75	65·65	69·13	73·29	89·33	107·6	113·1	118·1	124·1	128·3
100	67·33	70·06	74·22	77·93	82·36	99·33	118·5	124·3	129·6	135·8	140·2

Table A.2 Critical values of u in the runs test for n_1 or n_2 from 2 to 20. Any calculated value of u equal to or smaller than the values given in (a), or equal to or larger than those given in (b), is significant at the 0·05 level. (From Siegel 1956, pp. 252–3.)

(a) Values of u_L

n_1 \ n_2	2	3	4	5	6	7	8	9	10	11	12	13	14	15	16	17	18	19	20
2											2	2	2	2	2	2	2	2	2
3					2	2	2	2	2	2	2	2	2	3	3	3	3	3	3
4				2	2	2	3	3	3	3	3	3	3	3	4	4	4	4	4
5			2	2	3	3	3	3	3	4	4	4	4	4	4	4	5	5	5
6		2	2	3	3	3	3	4	4	4	4	5	5	5	5	5	5	6	6
7		2	2	3	3	3	4	4	5	5	5	5	5	6	6	6	6	6	6
8		2	3	3	3	4	4	5	5	5	6	6	6	6	6	7	7	7	7
9		2	3	3	4	4	5	5	5	6	6	6	7	7	7	7	8	8	8
10		2	3	3	4	5	5	5	6	6	7	7	7	7	8	8	8	8	9
11		2	3	4	4	5	5	6	6	7	7	7	8	8	8	9	9	9	9
12	2	2	3	4	4	5	6	6	7	7	7	8	8	8	9	9	9	10	10
13	2	2	3	4	5	5	6	6	7	7	8	8	9	9	9	10	10	10	10
14	2	2	3	4	5	5	6	7	7	8	8	9	9	9	10	10	10	11	11
15	2	3	3	4	5	6	6	7	7	8	8	9	9	10	10	11	11	11	12
16	2	3	4	4	5	6	6	7	8	8	9	9	10	10	11	11	11	12	12
17	2	3	4	4	5	6	7	7	8	9	9	10	10	11	11	11	12	12	13
18	2	3	4	5	5	6	7	8	8	9	9	10	10	11	11	12	12	13	13
19	2	3	4	5	6	6	7	8	8	9	10	10	11	11	12	12	13	13	13
20	2	3	4	5	6	6	7	8	9	9	10	10	11	12	12	13	13	13	14

(b) Values of u_U

n_1 \ n_2	2	3	4	5	6	7	8	9	10	11	12	13	14	15	16	17	18	19	20
2																			
3																			
4				9	9														
5			9	10	10	11	11												
6			9	10	11	12	12	13	13	13	13								
7				11	12	13	13	14	14	14	14	15	15	15					
8				11	12	13	14	14	15	15	16	16	16	17	17	17	17	17	
9					13	14	14	15	16	16	16	17	17	18	18	18	18	18	18
10					13	14	15	16	16	17	17	18	18	18	19	19	19	20	20
11					13	14	15	16	17	17	18	19	19	19	20	20	20	21	21
12					13	14	16	16	17	18	19	19	20	20	20	21	21	22	22
13						15	16	17	18	19	19	20	20	21	21	22	22	23	23
14						15	16	17	18	19	20	20	21	22	22	23	23	23	24
15						15	16	18	18	19	20	21	22	22	23	23	24	24	25
16							17	18	19	20	21	22	22	23	23	24	25	25	25
17							17	18	19	20	21	22	23	23	24	25	25	26	26
18							17	18	19	20	21	22	23	24	25	25	26	26	27
19							17	18	20	21	22	23	23	24	25	26	26	27	27
20							17	18	20	21	22	23	24	25	25	26	27	27	28

Table A.3   Areas under the normal curve. (From Weatherburn 1962, p. 56.)

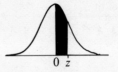

z	0·00	0·01	0·02	0·03	0·04	0·05	0·06	0·07	0·08	0·09
0·0	·0000	·0040	·0080	·0120	·0159	·0199	·0239	·0279	·0319	·0359
0·1	·0398	·0438	·0478	·0517	·0557	·0596	·0636	·0675	·0714	·0753
0·2	·0793	·0832	·0871	·0910	·0948	·0987	·1026	·1064	·1103	·1141
0·3	·1179	·1217	·1255	·1293	·1331	·1368	·1406	·1443	·1480	·1517
0·4	·1554	·1591	·1628	·1664	·1700	·1736	·1772	·1808	·1844	·1879
0·5	·1915	·1950	·1985	·2019	·2054	·2088	·2123	·2157	·2190	·2224
0·6	·2257	·2291	·2324	·2357	·2389	·2422	·2454	·2486	·2518	·2549
0·7	·2580	·2611	·2642	·2673	·2704	·2734	·2764	·2794	·2823	·2852
0·8	·2881	·2910	·2939	·2967	·2995	·3023	·3051	·3078	·3106	·3133
0·9	·3159	·3186	·3212	·3238	·3264	·3289	·3315	·3340	·3365	·3389
1·0	·3413	·3438	·3461	·3485	·3508	·3531	·3554	·3577	·3599	·3621
1·1	·3643	·3665	·3686	·3708	·3729	·3749	·3770	·3790	·3810	·3830
1·2	·3849	·3869	·3888	·3907	·3925	·3944	·3962	·3980	·3997	·4015
1·3	·4032	·4049	·4066	·4082	·4099	·4115	·4131	·4147	·4162	·4177
1·4	·4192	·4207	·4222	·4236	·4251	·4265	·4279	·4292	·4306	·4319
1·5	·4332	·4345	·4357	·4370	·4382	·4394	·4406	·4418	·4430	·4441
1·6	·4452	·4463	·4474	·4485	·4495	·4505	·4515	·4525	·4535	·4545
1·7	·4554	·4564	·4573	·4582	·4591	·4599	·4608	·4616	·4625	·4633
1·8	·4641	·4649	·4656	·4664	·4671	·4678	·4686	·4693	·4699	·4706
1·9	·4713	·4719	·4726	·4732	·4738	·4744	·4750	·4756	·4762	·4767
2·0	·4772	·4778	·4783	·4788	·4793	·4798	·4803	·4808	·4812	·4817
2·1	·4821	·4826	·4830	·4834	·4838	·4842	·4846	·4850	·4854	·4857
2·2	·4861	·4865	·4868	·4871	·4875	·4878	·4881	·4884	·4887	·4890
2·3	·4893	·4896	·4898	·4901	·4904	·4906	·4909	·4911	·4913	·4916
2·4	·4918	·4920	·4922	·4925	·4927	·4929	·4931	·4932	·4934	·4936
2·5	·4938	·4940	·4941	·4943	·4945	·4946	·4948	·4949	·4951	·4952
2·6	·4953	·4955	·4956	·4957	·4959	·4960	·4961	·4962	·4963	·4964
2·7	·4965	·4966	·4967	·4968	·4969	·4970	·4971	·4972	·4973	·4974
2·8	·4974	·4975	·4976	·4977	·4977	·4978	·4979	·4980	·4980	·4981
2·9	·4981	·4982	·4983	·4983	·4984	·4984	·4985	·4985	·4986	·4986
3·0	·49865	·4987	·4987	·4988	·4988	·4989	·4989	·4989	·4990	·4990
3·1	·49903	·4991	·4991	·4991	·4992	·4992	·4992	·4992	·4993	·4993

Table A.4   The Student *t* distribution.

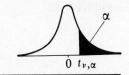

ν \ α	0·200	0·100	0·050	0·025	0·010	0·005
1	1·376	3·078	6·314	12·71	31·82	63·66
2	1·061	1·886	2·920	4·303	6·965	9·925
3	0·978	1·638	2·353	3·182	4·541	5·841
4	0·941	1·533	2·132	2·776	3·747	4·604
5	0·920	1·476	2·015	2·571	3·365	4·032
6	0·906	1·440	1·943	2·447	3·143	3·707
7	0·896	1·415	1·895	2·365	2·998	3·499
8	0·889	1·397	1·860	2·306	2·896	3·355
9	0·883	1·383	1·833	2·262	2·821	3·250
10	0·879	1·372	1·812	2·228	2·764	3·169
11	0·876	1·363	1·796	2·201	2·718	3·106
12	0·873	1·356	1·782	2·179	2·681	3·055
13	0·870	1·350	1·771	2·160	2·650	3·012
14	0·868	1·345	1·761	2·145	2·624	2·977
15	0·866	1·341	1·753	2·131	2·602	2·947
16	0·865	1·337	1·746	2·120	2·583	2·921
17	0·863	1·333	1·740	2·110	2·567	2·898
18	0·862	1·330	1·734	2·101	2·552	2·878
19	0·861	1·328	1·729	2·093	2·539	2·861
20	0·860	1·325	1·725	2·086	2·528	2·845
21	0·859	1·323	1·721	2·080	2·518	2·831
22	0·858	1·321	1·717	2·074	2·508	2·819
23	0·858	1·319	1·714	2·069	2·500	2·807
24	0·857	1·318	1·711	2·064	2·492	2·797
25	0·856	1·316	1·708	2·060	2·485	2·787
26	0·856	1·315	1·706	2·056	2·479	2·779
27	0·855	1·314	1·703	2·052	2·473	2·771
28	0·855	1·313	1·701	2·048	2·467	2·763
29	0·854	1·311	1·699	2·045	2·462	2·756
30	0·854	1.310	1.697	2·042	2·457	2·750
31	0·853	1·309	1·696	2·040	2·453	2·744
32	0·853	1·309	1·694	2·037	2·449	2·738
33	0·853	1·308	1·692	2·035	2·445	2·733
34	0·852	1·307	1·691	2·032	2·441	2·728
35	0·852	1·306	1·690	2·030	2·438	2·724
36	0·852	1·306	1·688	2·028	2·434	2·719
37	0·851	1·305	1·687	2·026	2·431	2·715
38	0·851	1·304	1·686	2·024	2·429	2·712
39	0·851	1·304	1·685	2·023	2·426	2·708
40	0·851	1·303	1·684	2·021	2·423	2·704
45	0·850	1·301	1·679	2·014	2·412	2·690
50	0·849	1·299	1·676	2·009	2·403	2·678
60	0·848	1·296	1·671	2·000	2·390	2·660
70	0·847	1·294	1·667	1·994	2·381	2·648
80	0·846	1·292	1·664	1·990	2·374	2·639
90	0·846	1·291	1·662	1·987	2·368	2·632
100	0·845	1·290	1·660	1·984	2·364	2·626
∞	0·842	1·282	1·645	1·960	2·326	2·576

Table A.5   The *F* distribution.

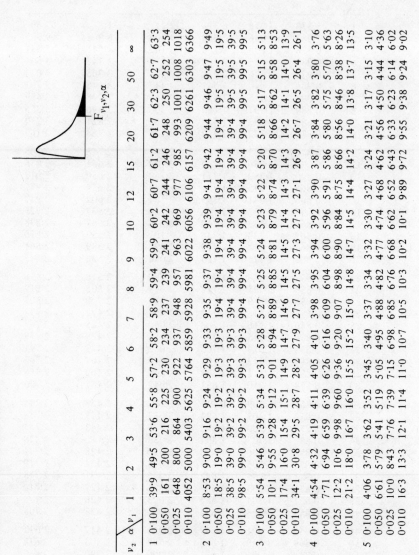

v_2	$\alpha \backslash v_1$	1	2	3	4	5	6	7	8	9	10	12	15	20	30	50	∞
1	0·100	39·9	49·5	53·6	55·8	57·2	58·2	58·9	59·4	59·9	60·2	60·7	61·2	61·7	62·3	62·7	63·3
	0·050	161	200	216	225	230	234	237	239	241	242	244	246	248	250	252	254
	0·025	648	800	864	900	922	937	948	957	963	969	977	985	993	1001	1008	1018
	0·010	4052	5000	5403	5625	5764	5859	5928	5981	6022	6056	6106	6157	6209	6261	6303	6366
2	0·100	8·53	9·00	9·16	9·24	9·29	9·33	9·35	9·37	9·38	9·39	9·41	9·42	9·44	9·46	9·47	9·49
	0·050	18·5	19·0	19·2	19·2	19·3	19·3	19·4	19·4	19·4	19·4	19·4	19·4	19·4	19·5	19·5	19·5
	0·025	38·5	39·0	39·2	39·2	39·3	39·3	39·4	39·4	39·4	39·4	39·4	39·4	39·4	39·5	39·5	39·5
	0·010	98·5	99·0	99·2	99·2	99·3	99·3	99·4	99·4	99·4	99·4	99·4	99·4	99·4	99·5	99·5	99·5
3	0·100	5·54	5·46	5·39	5·34	5·31	5·28	5·27	5·25	5·24	5·23	5·22	5·20	5·18	5·17	5·15	5·13
	0·050	10·1	9·55	9·28	9·12	9·01	8·94	8·89	8·85	8·81	8·79	8·74	8·70	8·66	8·62	8·58	8·53
	0·025	17·4	16·0	15·4	15·1	14·9	14·7	14·6	14·5	14·5	14·4	14·3	14·3	14·2	14·1	14·0	13·9
	0·010	34·1	30·8	29·5	28·7	28·2	27·9	27·7	27·5	27·3	27·2	27·1	26·9	26·7	26·5	26·4	26·1
4	0·100	4·54	4·32	4·19	4·11	4·05	4·01	3·98	3·95	3·94	3·92	3·90	3·87	3·84	3·82	3·80	3·76
	0·050	7·71	6·94	6·59	6·39	6·26	6·16	6·09	6·04	6·00	5·96	5·91	5·86	5·80	5·75	5·70	5·63
	0·025	12·2	10·6	9·98	9·60	9·36	9·20	9·07	8·98	8·90	8·84	8·75	8·66	8·56	8·46	8·38	8·26
	0·010	21·2	18·0	16·7	16·0	15·5	15·2	15·0	14·8	14·7	14·5	14·4	14·2	14·0	13·8	13·7	13·5
5	0·100	4·06	3·78	3·62	3·52	3·45	3·40	3·37	3·34	3·32	3·30	3·27	3·24	3·21	3·17	3·15	3·10
	0·050	6·61	5·79	5·41	5·19	5·05	4·95	4·88	4·82	4·77	4·74	4·68	4·62	4·56	4·50	4·44	4·36
	0·025	10·0	8·43	7·76	7·39	7·15	6·98	6·85	6·76	6·68	6·62	6·52	6·43	6·33	6·23	6·14	6·02
	0·010	16·3	13·3	12·1	11·4	11·0	10·7	10·5	10·3	10·2	10·1	9·89	9·72	9·55	9·38	9·24	9·02

$F_{v_1 v_2 \alpha}$

6	0·100	3·78	3·46	3·29	3·18	3·11	3·05	3·01	2·98	2·96	2·94	2·90	2·87	2·84	2·80	2·77	2·72
	0·050	5·99	5·14	4·76	4·53	4·39	4·28	4·21	4·15	4·10	4·06	4·00	3·94	3·87	3·81	3·75	3·67
	0·025	8·81	7·26	6·60	6·23	5·99	5·82	5·70	5·60	5·52	5·46	5·37	5·27	5·17	5·07	4·98	4·85
	0·010	13·7	10·9	9·78	9·15	8·75	8·47	8·26	8·10	7·98	7·87	7·72	7·56	7·40	7·23	7·09	6·88
7	0·100	3·59	3·26	3·07	2·96	2·88	2·83	2·78	2·75	2·72	2·70	2·67	2·63	2·59	2·56	2·52	2·47
	0·050	5·59	4·74	4·35	4·12	3·97	3·87	3·79	3·73	3·68	3·64	3·57	3·51	3·44	3·38	3·32	3·23
	0·025	8·07	6·54	5·89	5·52	5·29	5·12	4·99	4·90	4·82	4·76	4·67	4·57	4·47	4·36	4·28	4·14
	0·010	12·2	9·55	8·45	7·85	7·46	7·19	6·99	6·84	6·72	6·62	6·47	6·31	6·16	5·99	5·86	5·65
8	0·100	3·46	3·11	2·92	2·81	2·73	2·67	2·62	2·59	2·56	2·54	2·50	2·46	2·42	2·38	2·35	2·29
	0·050	5·32	4·46	4·07	3·84	3·69	3·58	3·50	3·44	3·39	3·35	3·28	3·22	3·15	3·08	3·02	2·93
	0·025	7·57	6·06	5·42	5·05	4·82	4·65	4·53	4·43	4·36	4·30	4·20	4·10	4·00	3·89	3·81	3·67
	0·010	11·3	8·65	7·59	7·01	6·63	6·37	6·18	6·03	5·91	5·81	5·67	5·52	5·36	5·20	5·07	4·86
9	0·100	3·36	3·01	2·81	2·69	2·61	2·55	2·51	2·47	2·44	2·42	2·38	2·34	2·30	2·25	2·22	2·16
	0·050	5·12	4·26	3·86	3·63	3·48	3·37	3·29	3·23	3·18	3·14	3·07	3·01	2·94	2·86	2·80	2·71
	0·025	7·21	5·71	5·08	4·72	4·48	4·32	4·20	4·10	4·03	3·96	3·87	3·77	3·67	3·56	3·47	3·33
	0·010	10·6	8·02	6·99	6·42	6·06	5·80	5·61	5·47	5·35	5·26	5·11	4·96	4·81	4·65	4·52	4·31
10	0·100	3·29	2·92	2·73	2·61	2·52	2·46	2·41	2·38	2·35	2·32	2·28	2·24	2·20	2·16	2·12	2·06
	0·050	4·96	4·10	3·71	3·48	3·33	3·22	3·14	3·07	3·02	2·98	2·91	2·85	2·77	2·70	2·64	2·54
	0·025	6·94	5·46	4·83	4·47	4·24	4·07	3·95	3·85	3·78	3·72	3·62	3·52	3·42	3·31	3·22	3·08
	0·010	10·0	7·56	6·55	5·99	5·64	5·39	5·20	5·06	4·94	4·85	4·71	4·56	4·41	4·25	4·12	3·91
12	0·100	3·18	2·81	2·61	2·48	2·39	2·33	2·28	2·24	2·21	2·19	2·15	2·10	2·06	2·01	1·97	1·90
	0·050	4·75	3·89	3·49	3·26	3·11	3·00	2·91	2·85	2·80	2·75	2·69	2·62	2·54	2·47	2·40	2·30
	0·025	6·55	5·10	4·47	4·12	3·89	3·73	3·61	3·51	3·44	3·37	3·28	3·18	3·07	2·96	2·87	2·72
	0·010	9·33	6·93	5·95	5·41	5·06	4·82	4·64	4·50	4·39	4·30	4·16	4·01	3·86	3·70	3·57	3·36
14	0·100	3·10	2·73	2·52	2·39	2·31	2·24	2·19	2·15	2·12	2·10	2·05	2·01	1·96	1·91	1·87	1·80
	0·050	4·60	3·74	3·34	3·11	2·96	2·85	2·76	2·70	2·65	2·60	2·53	2·46	2·39	2·31	2·24	2·13
	0·025	6·30	4·86	4·24	3·89	3·66	3·50	3·38	3·29	3·21	3·15	3·05	2·95	2·84	2·73	2·64	2·49
	0·010	8·86	6·51	5·56	5·04	4·69	4·46	4·28	4·14	4·03	3·94	3·80	3·66	3·51	3·35	3·22	3·00
16	0·100	3·05	2·67	2·46	2·33	2·24	2·18	2·13	2·09	2·06	2·03	1·99	1·94	1·89	1·84	1·79	1·72
	0·050	4·49	3·63	3·24	3·01	2·85	2·74	2·66	2·59	2·54	2·49	2·42	2·35	2·28	2·19	2·12	2·01
	0·025	6·12	4·69	4·08	3·73	3·50	3·34	3·22	3·12	3·05	2·99	2·89	2·79	2·68	2·57	2·47	2·32
	0·010	8·53	6·23	5·29	4·77	4·44	4·20	4·03	3·89	3·78	3·69	3·55	3·41	3·26	3·10	2·97	2·75

Table A.5   (continued)

v_2	α	1	2	3	4	5	6	7	8	9	10	12	15	20	30	50	∞
18	0·100	3·01	2·62	2·42	2·29	2·20	2·13	2·08	2·04	2·00	1·98	1·93	1·89	1·84	1·78	1·74	1·66
	0·050	4·41	3·55	3·16	2·93	2·77	2·66	2·58	2·51	2·46	2·41	2·34	2·27	2·19	2·11	2·04	1·92
	0·025	5·98	4·56	3·95	3·61	3·38	3·22	3·10	3·01	2·93	2·87	2·77	2·67	2·56	2·44	2·35	2·19
	0·010	8·29	6·01	5·09	4·58	4·25	4·01	3·84	3·71	3·60	3·51	3·37	3·23	3·08	2·92	2·78	2·57
20	0·100	2·97	2·59	2·38	2·25	2·16	2·09	2·04	2·00	1·96	1·94	1·89	1·84	1·79	1·74	1·69	1·61
	0·050	4·35	3·49	3·10	2·87	2·71	2·60	2·51	2·45	2·39	2·35	2·28	2·20	2·12	2·04	1·97	1·84
	0·025	5·87	4·46	3·86	3·51	3·29	3·13	3·01	2·91	2·84	2·77	2·68	2·57	2·46	2·35	2·25	2·09
	0·010	8·10	5·85	4·94	4·43	4·10	3·87	3·70	3·56	3·46	3·37	3·23	3·09	2·94	2·78	2·64	2·42
25	0·100	2·92	2·53	2·32	2·18	2·09	2·02	1·97	1·93	1·89	1·87	1·82	1·77	1·72	1·66	1·61	1·52
	0·050	4·24	3·39	2·99	2·76	2·60	2·49	2·40	2·34	2·28	2·24	2·16	2·09	2·01	1·92	1·84	1·71
	0·025	5·69	4·29	3·69	3·35	3·13	2·97	2·85	2·75	2·68	2·61	2·51	2·41	2·30	2·18	2·08	1·91
	0·010	7·77	5·57	4·68	4·18	3·85	3·63	3·46	3·32	3·22	3·13	2·99	2·85	2·70	2·54	2·40	2·17
30	0·100	2·88	2·49	2·28	2·14	2·05	1·98	1·93	1·88	1·85	1·82	1·77	1·72	1·67	1·61	1·55	1·46
	0·050	4·17	3·32	2·92	2·69	2·53	2·42	2·33	2·27	2·21	2·16	2·09	2·01	1·93	1·84	1·76	1·62
	0·025	5·57	4·18	3·59	3·25	3·03	2·87	2·75	2·65	2·57	2·51	2·41	2·31	2·20	2·07	1·97	1·79
	0·010	7·56	5·39	4·51	4·02	3·70	3·47	3·30	3·17	3·07	2·98	2·84	2·70	2·55	2·39	2·25	2·01
40	0·100	2·84	2·44	2·23	2·09	2·00	1·93	1·87	1·83	1·79	1·76	1·71	1·66	1·61	1·54	1·48	1·38
	0·050	4·08	3·23	2·84	2·61	2·45	2·34	2·25	2·18	2·12	2·08	2·00	1·92	1·84	1·74	1·66	1·51
	0·025	5·42	4·05	3·46	3·13	2·90	2·74	2·62	2·53	2·45	2·39	2·29	2·18	2·07	1·94	1·83	1·64
	0·010	7·31	5·18	4·31	3·83	3·51	3·29	3·12	2·99	2·89	2·80	2·66	2·52	2·37	2·20	2·06	1·80
50	0·100	2·81	2·41	2·20	2·06	1·97	1·90	1·84	1·80	1·76	1·73	1·68	1·63	1·57	1·50	1·44	1·33
	0·050	4·03	3·18	2·79	2·56	2·40	2·29	2·20	2·13	2·07	2·03	1·95	1·87	1·78	1·69	1·60	1·44
	0·025	5·34	3·97	3·39	3·05	2·83	2·67	2·55	2·46	2·38	2·32	2·22	2·11	1·99	1·87	1·75	1·55
	0·010	7·17	5·06	4·20	3·72	3·41	3·19	3·02	2·89	2·78	2·70	2·56	2·42	2·27	2·10	1·95	1·68
60	0·100	2·79	2·39	2·18	2·04	1·95	1·87	1·82	1·77	1·74	1·71	1·66	1·60	1·54	1·48	1·41	1·29
	0·050	4·00	3·15	2·76	2·53	2·37	2·25	2·17	2·10	2·04	1·99	1·92	1·84	1·75	1·65	1·56	1·39
	0·025	5·29	3·93	3·34	3·01	2·79	2·63	2·51	2·41	2·33	2·27	2·17	2·06	1·94	1·82	1·70	1·48
	0·010	7·08	4·98	4·13	3·65	3·34	3·12	2·95	2·82	2·72	2·63	2·50	2·35	2·20	2·03	1·88	1·60

Table A.6 Critical values of U for the Mann–Whitney test. (From Siegel 1956, pp. 275–6.)

(a) Critical values for one-tailed test at $\alpha = 0.025$, or for two-tailed test at $\alpha = 0.05$.

n_1 \ n_2	9	10	11	12	13	14	15	16	17	18	19	20
1												
2	0	0	0	1	1	1	1	1	2	2	2	2
3	2	3	3	4	4	5	5	6	6	7	7	8
4	4	5	6	7	8	9	10	11	11	12	13	13
5	7	8	9	11	12	13	14	15	17	18	19	20
6	10	11	13	14	16	17	19	21	22	24	25	27
7	12	14	16	18	20	22	24	26	28	30	32	34
8	15	17	19	22	24	26	29	31	34	36	38	41
9	17	20	23	26	28	31	34	37	39	42	45	48
10	20	23	26	29	33	36	39	42	45	48	52	55
11	23	26	30	33	37	40	44	47	51	55	58	62
12	26	29	33	37	41	45	49	53	57	61	65	69
13	28	33	37	41	45	50	54	59	63	67	72	76
14	31	36	40	45	50	55	59	64	67	74	78	83
15	34	39	44	49	54	59	64	70	75	80	85	90
16	37	42	47	53	59	64	70	75	81	86	92	98
17	39	45	51	57	63	67	75	81	87	93	99	105
18	42	48	55	61	67	74	80	86	93	99	106	112
19	45	52	58	65	72	78	85	92	99	106	113	119
20	48	55	62	69	76	83	90	98	105	112	119	127

(b) Critical values for one-tailed test at $\alpha = 0.01$, or for two-tailed test at $\alpha = 0.02$.

n_1 \ n_2	9	10	11	12	13	14	15	16	17	18	19	20
1												
2					0	0	0	0	0	0	1	1
3	1	1	1	2	2	2	3	3	4	4	4	5
4	3	3	4	5	5	6	7	7	8	9	9	10
5	5	6	7	8	9	10	11	12	13	14	15	16
6	7	8	9	11	12	13	15	16	18	19	20	22
7	9	11	12	14	16	17	19	21	23	24	26	28
8	11	13	15	17	20	22	24	26	28	30	32	34
9	14	16	18	21	23	26	28	31	33	36	38	40
10	16	19	22	24	27	30	33	36	38	41	44	47
11	18	22	25	28	31	34	37	41	44	47	50	53
12	21	24	28	31	35	38	42	46	49	53	56	60
13	23	27	31	35	39	43	47	51	55	59	63	67
14	26	30	34	38	43	47	51	56	60	65	69	73
15	28	33	37	42	47	51	56	61	66	70	75	80
16	31	36	41	46	51	56	61	66	71	76	82	87
17	33	38	44	49	55	60	66	71	77	82	88	93
18	36	41	47	53	59	65	70	76	82	88	94	100
19	38	44	50	56	63	69	75	82	88	94	101	107
20	40	47	53	60	67	73	80	87	93	100	107	114

266 APPENDICES

Table A.7   Random numbers (from Neave 1978, p. 64).

68327	00067	17487	49149	25894	23639	86557	04139	10756	76285
55888	82253	67464	91628	88764	43598	45481	00331	15900	97699
84910	44827	31173	44247	56573	91759	79931	26644	27048	53704
35654	53638	00563	57230	07395	10813	99194	81592	96834	21374
46381	60071	20835	43110	31842	02855	73446	24456	24268	85291
11212	06034	77313	66896	47902	63483	09924	83635	30013	61791
49703	07226	73337	49223	73312	09534	64005	79267	76590	26066
05482	30340	24606	99042	16536	14267	84084	16198	94852	44305
92947	65090	47455	90675	89921	13036	92867	04786	76776	18675
51806	61445	32437	01129	03644	70024	07629	55805	85616	59569
16383	30577	91319	67998	72423	81307	75192	80443	09651	30068
30893	85406	42369	71836	74479	68273	78133	34506	68711	58725
59790	11682	63156	10443	99033	76460	36814	36917	37232	66218
06271	74980	46094	21881	43525	16516	26393	89082	24343	57546
93325	61834	40763	81178	17507	90432	50973	35591	36930	03184
46690	08927	32962	24882	83156	58597	88267	32479	80440	41668
82041	88942	57572	34539	43812	58483	43779	42718	46798	49079
14306	04003	91186	70093	62700	99408	72236	52722	37531	24590
63471	77583	80056	59027	37031	05819	90836	19530	07138	36431
68467	17634	84211	31776	92996	75644	82043	84157	10877	12536
94308	57895	08121	07088	65080	51928	74237	00449	86625	06626
52218	32502	82195	43867	79935	34620	37386	00243	46353	44499
46586	08309	52702	85464	06670	18796	74713	81632	34056	56461
07869	80471	69139	82408	33989	44250	79597	15182	14956	70423
46719	60281	88638	26909	32415	31864	53708	60219	44482	40004
74687	71227	59716	80619	56816	73807	94150	21991	22901	74351
42731	50249	11685	54034	12710	35159	00214	19440	61539	25717
71740	29429	86822	01187	96497	25823	18415	06087	05886	11205
96746	05938	11828	47727	02522	33147	92846	15010	96725	67903
27564	81744	51909	36192	45263	33212	71808	24753	72644	74441
21895	29683	26533	14740	94286	90342	24674	52762	22051	31743
01492	40778	05988	65760	13468	31132	37106	02723	40202	15824
55846	19271	22846	80425	00235	34292	72181	24910	25245	81239
14615	75196	40313	50783	66585	39010	76796	31385	26785	66830
77848	15755	91938	81915	65312	86956	26195	61525	97406	67988
87167	03106	52876	31670	23850	13257	77510	42393	53782	32412
73018	56511	89388	73133	12074	62538	57215	23476	92150	14737
29247	67792	10593	22772	03407	24319	19525	24672	21182	10765
17412	09161	34905	44524	20124	85151	25952	81930	43536	39705
68805	19830	87973	99691	25096	41497	57562	35553	77057	06161
40551	36740	61851	76158	35441	66188	87728	66375	98049	84604
90379	06314	21897	42800	63963	44258	14381	90884	66620	14538
09466	65311	95514	51559	29960	07521	42180	86677	94240	59783
15821	25078	19388	93798	50820	88254	20504	74158	35756	42100
10328	60890	05204	30069	79630	31572	63273	13703	52954	72793
49727	08160	81650	71690	56327	06729	22495	49756	43333	34533
71118	41798	34541	76432	40522	51521	74382	06305	11956	30611
53253	23100	03743	48999	37736	92186	19108	69017	21661	17175
12206	24205	32372	46438	67981	53226	24943	68659	91924	69555

REFERENCES TO STATISTICAL TABLES

Tables A.1, A.4, A.5 and A.7 are taken from:
Neave, H. R. 1978. *Statistics tables*. London: George Allen & Unwin.
Other tables are from:
Siegel, S. 1956. *Nonparametric statistics for the behavioral sciences*. New York: McGraw-Hill.
Weatherburn, C. E. 1962. *A first course in mathematical statistics*. Cambridge: Cambridge University Press.

GLOSSARY OF SYMBOLS

Symbols appearing in the text are defined below, followed by the page number on which they first occur.

Mathematical operators are listed first. English letters and Greek letters are listed separately in their approximate alphabetical order.

MATHEMATICAL OPERATORS

Symbol	Definition	Page
$\neq$	not equal to	103
$\approx$	approximately equal to	100
$a < b$	a is less than b	68
$a > b$	a is greater than b	68
$\leqslant$	less than or equal to	39
$\geqslant$	greater than or equal to	87
$\sqrt{}$	square root	19
X^a	X raised to the ath power	76
X^{-a}	X raised to the minus ath power. N.B. $X^{-a} = 1/X^a$	100
$\log_e$	logarithm to the base e (natural logarithm): $\log_e e = 1$, $\log_e e^2 = 2$ etc.	241
$\log_{10}$	logarithm to the base 10 (common logarithm): $\log_{10} 10 = 1$, $\log_{10} 10^2 = 2$, etc.	12
$n!$	n factorial: multiply n by all whole numbers less than it but greater than zero: $$n(n-1)(n-2)\ldots(2)(1)$$	76
$\binom{n}{k}$	binomial coefficient = calculated from $$\frac{n!}{(n-k)!\,k!}$$	76
$\lvert X \rvert$	absolute value of X: the value of X ignoring its sign	22
Σ	sum all quantities or scores that follow	16
$\sum\limits_{i=1}^{n} X_i$	sum all quantities X_1 through X_n: $$X_1 + X_2 + \cdots + X_n$$	16

Operations with logarithms (natural or common)

$\log(ab) = \log a + \log b$	240
$\log(a/b) = \log a - \log b$	240
$\log(X^a) = a \log X$	240
$\log(X^{-a}) = -a \log X$	240

GREEK LETTERS

Symbol	Definition	Page
α	(1) Probability of a type 1 error, probability of rejecting H_0 when it is true	51
	(2) Parameter representing the true intercept term of a regression equation	229
$\hat{\alpha}$	least-squares estimate of $\alpha(2)$	232
β	parameter representing the true slope term of a regression equation	229
$\hat{\beta}$	least-squares estimate of β	232
χ^2	chi square	46
δ	mean first nearest neighbour distance expected under the null hypothesis of randomness	107
ϵ_i	population error or disturbance term in the linear regression model	229
λ	density of points per quadrat or per unit area in point pattern analysis	100
μ	population mean	134
$\mu_{1-\alpha}$	$(1-\alpha) \times 100\%$ confidence interval for μ	158
ν	degrees of freedom	46
ϕ	phi coefficient, or contingency coefficient	200
π	ratio of the circumference of a circle to its diameter. Its numerical value is $3\cdot14159\ldots$	84
Π	population proportion	134
$\Pi_{1-\alpha}$	$(1-\alpha) \times 100\%$ confidence interval for Π	157
ρ	population correlation coefficient	209
ρ_S	Spearman's rank population correlation coefficient	202
σ	population standard deviation	19 (134)
σ^2	population variance	18 (134)
σ_p	true standard error of a sample proportion p	151
σ_R	standard deviation of the residuals about regression	226
$\sigma_{\bar{X}}$	true standard error of the mean	158

ENGLISH LETTERS

Symbol	Definition	Page
a	intercept or constant term in a regression equation	219
b	slope term in a regression equation	219
cv	coefficient of variation based on sample data	233
CV	coefficient of variation based on population data	20
C_j	midpoint of jth class	17
d_{ij}	distance between two items or individuals located at the points (U_i, V_i) and (U_j, V_j)	107
$\bar{d}$	observed mean first nearest neighbour distance	107
df	degrees of freedom	48
e	the base of the natural system of logarithms. Its numerical value is $2\cdot71828\ldots.$	84

Symbol	Definition	Page
e_i	(1) expected number in a class or category (2) $= Y_i - \hat{Y}_i$. Residual, difference between observed and estimated values of Y in regression analysis	46 225
$E(X)$	expected value of X	48 (14
f_j	frequency (number) of observations in jth class	12
F	a ratio of two variances	172
H	statistic in the Kruskall–Wallis analysis of variance by ranks	193
H_0	the null hypothesis	50
H_1	the alternative hypothesis	50
n	number of items or individuals	9
n_i	number of items, individuals in ith group	50
o_i	observed number in a class or category	46
p	(1) sample proportion (2) probability of occurrence of an event	134 73
$P(A)$	probability of event A	39
$P(B/A)$	conditional probability; probability of B given that A has occurred	45
q	probability of non-occurrence of an event	73
r	Pearson product-moment correlation coefficient	203
r_S	Spearman's rank correlation coefficient	202
R^2	coefficient of determination	223
s^2	sample variance	134
s	sample standard deviation	134
s_D	standard distance deviation	24
s_p	estimated standard error of a sample proportion p	154
$s^2_{p_1 - p_2}$	estimated variance of the difference between two sample proportions, p_1 and p_2	168
s_r	estimated standard error of the sample correlation coefficient, r	211
s_R	standard error of estimate (about regression)	233
$s^2_{\bar{X}}$	estimated variance of the sampling distribution of the mean	158
$s_{\bar{X}}$	estimated standard error of the mean	158
$s_{\bar{X}_1 - \bar{X}_2}$	estimated standard error of the difference between two sample means	171
s_Z	estimated standard error of transformed value of sample correlation coefficient	212
t	statistic used to test hypotheses when sample size is 'small' $(n < 25)$	159
U_1, U_2	statistics in the Mann–Whitney U test	186
U_i	'easting' of location of ith individual or item	22
V_i	'northing' of location of ith individual or item	22
(U_i, V_i)	location of ith individual or item	22
$(\bar{U}, \bar{V})$	mean centre	23
X, Y	variables	9
X_i, Y_i	specific values denoted by the subscript i	9
$\bar{X}, \bar{Y}$	arithmetic means	16
$\hat{Y}_i$	value of Y, corresponding to $X = X_i$, estimated by a regression equation	221

Symbol	Definition	Page
z	(1) 'z score': deviation of a given score from the mean expressed in standard deviation units	86
	(2) statistic used to test hypotheses when sample size is sufficiently large ($n \geqslant 25$)	92 (158)
Z	transformed value of a sample correlation coefficient	212

Index